THE NUCLEAR IMPERATIVE

TOPICS IN SAFETY, RISK, RELIABILITY AND QUALITY

VOLUME 11

Editor
Adrian.V. Gheorghe
Swiss Federal Institute of Technology, Zürich, Switzerland

Editorial Advisory Board
P. Sander, *Technical University of Eindhoven, The Netherlands*
D.C. Barrie, *Lakehead University, Ontario, Canada*
R. Leitch, *Royal Military College of Science (Cranfield), Shriverham, U.K.*

Aims and Scope. Fundamental questions which are being asked these days of all products, processes and services with ever increasing frequency are:

> What is the risk?
> How safe is it?
> How reliable is it?
> How good is the quality?
> How much does it cost?

This is particularly true as the government, industry, public, customers and society become increasingly informed and articulate.

In practice none of the three topics can be considered in isolation as they all interact and interrelate in very complex and subtle ways and require a range of disciplines for their description and application; they encompass the social, engineering and physical sciences and quantitative disciplines including mathematics, probability theory and statistics.

The major objective of the series is to provide a series of authoritative texts suitable for academic taught courses, reference purposes, post graduate and other research and practitioners generally working or strongly associated with areas such as:

> Safety Assessment and Management
> Emergency Planning
> Risk Management
> Reliability Analysis and Assessment
> Vulnerability Assessment and Management
> Quality Assurance and Management

Special emphasis is placed on texts with regard to readability, relevance, clarity, applicability, rigour and generally sound quantitative content.

The titles published in this series are listed at the end of this volume.

The Nuclear Imperative

A Critical Look at the Approaching Energy Crisis

by

Jeff W. Eerkens
Nuclear Science and Engineering Institute
University of Missouri, Columbia, U.S.A.

Springer

A C.I.P. Catalogue record for this book is available from the Library of Congress.

ISBN-10 1-4020-4930-7 (HB)
ISBN-13 978-1-4020-4930-9 (HB)
ISBN-10 1-4020-4931-5 (e-book)
ISBN-13 978-1-4020-4931-6 (e-book)

Published by Springer,
P.O. Box 17, 3300 AA Dordrecht, The Netherlands.

www.springer.com

Printed on acid-free paper

All Rights Reserved
© 2006 Springer
No part of this work may be reproduced, stored in a retrieval system, or transmitted
in any form or by any means, electronic, mechanical, photocopying, microfilming, recording
or otherwise, without written permission from the Publisher, with the exception
of any material supplied specifically for the purpose of being entered
and executed on a computer system, for exclusive use by the purchaser of the work.

Printed in the Netherlands.

*This book is dedicated
to the memory of
pioneer nuclear physicist
Prof. Dr. Hans Bethe, 1906–2005*

'*He probed and deciphered the cosmos,
revealing the enormous energy
locked up in atomic nuclei*'

CONTENTS

Preface xi

List of Briefs xv

Abstract xvii

Chapter 1: Introduction 1

1.1. Declaration of Independence from Oil 1
1.2. Needed Actions and Risks to Overcome the Pending No-Oil Crisis 4
    1.2.1. Risks in the Nuclear Millennium Option 6
    1.2.2. Risks of the Renewables-Only Option 12
    1.2.3. The Coal-Burning Option 14
    1.2.4. The Do-Nothing Option 15
    1.2.5. Conclusions based on the Risk Assessment of Options (1), (2), (3), and (4) 16
1.3. Organization of the Book 18

Chapter 2: Nuclear Facts and Fables 19

Chapter 3: Energy Consumption and Energy Sources on Planet Earth 31

3.1. Definition of Energy and Units for Energy and Power 31
3.2. Amounts and Forms of Energy Consumed by Man 34
3.3. Limitations of 'Renewable Energy' 38
3.4. A Brief History of Energy 41
3.5. Summary of Primary Energy Sources 46

Chapter 4: Technologies for Propelling Cars, Trucks, Trains, Ships and Aircraft 51

4.1. Review of Portable Fuels and Other Energy Carriers 52
    4.1.1. Portable Synfuels 52
    4.1.2. Electric Storage Batteries 57
    4.1.3. Flywheel Energy Storage (Mechanical Batteries) 59
4.2. Vehicle Propulsion Engines 61
    4.2.1. Internal Combustion Engines (ICEs) 61

viii   CONTENTS

        4.2.2.   Electrochemical Fuel-Cell Engines (FCEs)   63
        4.2.3.   Steam Engine   67

Chapter 5: Electric Power Generation Technology   69

5.1. Nuclear Power Reactors   70
       5.1.1.   Basic Design and Operation of a Reactor   72
       5.1.2.   Breeder Reactors   77
       5.1.3.   Nuclear Reactors versus Nuclear Bombs   79
5.2. The Uranium Fuel Cycle and Environmental Impacts   81
       5.2.1.   Uranium Enrichment   84
       5.2.2.   Fuel (Re-)Processing   87
       5.2.3.   Final Disposal of Radiowaste   88
5.3. Nuclear Fusion   92

Chapter 6: Safety Considerations in Nuclear Operations   97

6.1. Nature of Nuclear Radiation   97
6.2. Biological Effects of Nuclear Radiation   99
6.3. Radiation Dose Measurements and Tolerable Exposures   104
6.4. Radiation Protection in Reactor Operations   108
6.5. Mishaps and Malfunctions in Reactor Operations   109
6.6. Nuclear Criticality Accidents   110

Chapter 7: Measures and Means to Control the Global Use of Nuclear Weapons   117

7.1. The Nuclear Age and World Realities   117
7.2. Safeguards against Terrorists   122
7.3. The Non-Proliferation Treaty (NPT) and International Atomic Energy Agency (IAEA)   125

Chapter 8: Conclusions, Action Items, and Predictions   131

8.1. Investigations of Synthetic Fuel (Synfuel) Manufacture, Storage, Distribution, and Selection   132
8.2. Development of Advanced Auto/Aircraft Engines and Other Energy supplying Devices   133
8.3. Nuclear Breeder Reactor Development, Testing, and Deployment   134
8.4. Coal Usage Reassignment Programs   134
8.5. Anticipations   135

Annotated Bibliography   139

Acknowledgements   143

Short Biography of the Author	**145**
Abbreviations	**147**
Chemical Symbols of Selected Elements and Isotopes	**149**
International MKS Units and Prefixes	**151**
Index	**153**

PREFACE

Part of this book was originally written in 1976 when some thirty copies were printed and distributed to interested parties. A wider distribution was planned but never carried out. In re-reading the first edition written 29 years ago, I am struck by the fact that arguments for a nuclear-powered future and a need to develop portable synfuels and new engines have remained unchanged over all these years. Nuclear proliferation and international weapon control issues in 1976 are also quite applicable still today.

This book can be used as a textbook in an introductory course on nuclear engineering. More generally, the subject matter can be reviewed in a first-year college class for students planning a career in engineering, economics, political science, law, and for those who will be involved with structuring our future society. The book is also recommended reading for high-school seniors contemplating a higher education. Some familiarity with high-school physics and chemistry is helpful, but 'one does not have to be a rocket scientist' to understand the essential issues. All material is based on 'hard' science, as opposed to dubious 'pop' or 'junk' science one often sees which distorts the facts and exploits people's predilection for the sensational.

The Three-Mile-Island (TMI) reactor meltdown in the US which occurred on March 28, 1979, and the Chernobyl reactor disaster that took place on April 26, 1986, caused an unfortunate slow-down and in some countries a moratorium on the construction of new nuclear power plants. However these accidents resulted in vastly improved safety measures in nuclear reactor operations. They also proved that a 'maximum credible' reactor accident (a reactor meltdown) does not 'kill thousands of people' as predicted by those opposed to the development of nuclear energy. In fact, the early safety measures built into American and West-European reactors, comprising a large steel and concrete containment vessel and other features, were shown to perform as designed. Thanks to TMI's containment vessel and pressurized water coolant system which provides a negative temperature coefficient of reactivity, the TMI accident in Pennsylvania did not harm a single person. However the Chernobyl reactor in the Ukraine had no containment vessel and used a graphite moderator with positive reactivity coefficient. Operator error induced a runaway reactor melt-down, causing the graphite to burn chemically with atmospheric oxygen. Non-nuclear-educated firemen from nearby cities who attempted to put out the fire, received serious overexposures of radiation. They had not been warned thereof due to bureaucratic territorialism and unnecessary secrecy. Ultimately 45 people died due to the Chernobyl reactor meltdown and 30 suffered permanent disabilities according to the International Atomic Energy Agency. Today's regulations require all nuclear power reactors to shut themselves down when they get too hot (= negative coefficient of reactivity) and to provide a containment vessel around the reactor, designed to hold all radioactive debris in case of a reactor melt-down accident.

While the electric power industry suffered a set-back in the public acceptance of nuclear energy, the reasons for expanding construction of new nuclear power plants three decades ago have not changed and in fact have become more urgent. After demise of oil and gaseous petrofuels from the earth, only coal and uranium (perhaps deuterium/tritium in the next century) are left as prime sources to provide us with large-scale quantities of energy. Large quantities of prime energy are needed in the future to synthesize hydrogen, ammonia, hydrazine, and other portable 'synfuels' to move our transportation fleets (cars, trucks, trains, ships, aircraft, etc.), and to energize our heavy industries (steel-production, ship-building, auto manufacturing, etc.). Uranium can satisfy this demand for at least 1000 years with much less waste and real estate problems and with vastly better economics than 'renewable' solar-cells and wind-farms ever could. In the USA, coal-fired power plants presently produce 52%, uranium fission generates 21%, natural gas and oil-burning units contribute 15%, and hydro + geothermal + others yield 12% of all electricity (3.5 billion MWh(e)/yr). Burning of coal, oil, or natural gas produces heat. In a power plant this heat vaporizes water into high-pressure steam which is converted into electricity by steam turbines. In a nuclear plant, water is heated by fissioning uranium; otherwise the same electric generation is used as in coal-fired plants, that is with steam turbines. However coal, oil and natural gas pollute the atmosphere enormously when burnt with atmospheric oxygen (O_2), since they produce globe-warming gaseous carbon dioxide (CO_2). In theory, without oil and uranium, coal could supply the world with all needed energy for some hundred years. But it would be insanity not to generate electricity for the next thousand years utilizing non-polluting fissioning of uranium instead of burning coal, knowing that uranium has little other use. When oil is gone, only coal (and wood) can still provide raw material for making plastics and other carbonaceous compounds widely used today; it should not be burnt! The oft-quoted 'problem of radioactive waste' produced by reactors is highly exaggerated. 'The problem' is non-existent for nuclear engineers who have no difficulty concentrating and packing solid oxidized fission products from one hundred US nuclear power plants in a few hundred drums yearly for underground storage.

This book is not the first nor will it be the last one that warns of an impending energy crisis. The annotated bibliography lists books giving similar serious warnings, which seem to be unheeded by mal-informed governments. The main message of this book is that doom of our civilization due to depletion of oil is not inevitable if the correct measures are taken. Current government energy policies seem influenced by dogmatically anti-nuclear lobbyists who believe only 'renewable' wind and solar farms can solve future energy problems. Nuclear and coal power plant experts should be consulted instead when making policy decisions. The sun provides a year-averaged 500 watts of light energy per square meter, and most strong winds blow only 20% of the time. No amount of legislation can alter those facts. Also in round numbers, the earth once possessed 5 trillion barrels of oil and 20 quadrillion cubic feet of natural gas (natgas) from fossil remains, including oil from tar-sands and natgas from sea-beds. Of these, 3.5 trillion barrels of oil and 15 quadrillion cubic feet of natgas are left today. In addition earth also possesses 6 trillion tons of coal and 10 million tons of exploitable uranium at present. With average energy consumption leveling to 1.3 kW(e) per person and a stabilizing world population of 7 billion people, it would take 17 years to deplete all oil, 18 years for all natgas, 153 years for all coal, and 1000 years to burn up all uranium, *assuming* each were the *only* prime energy source for electricity and portable fuel production. Presently (2005), world consumption is 0.7 kW(e) per person, but Asia is expected to raise this to 1.3 kW(e) in ten years. The U.S., with 4.5% of the world population, consumes 4.4 kW(e) per capita at present. However roughly 1.2 kW(e) is spent on the electricity-intensive manufacture of exported goods (aircraft, autos, bulldozers, ships, etc) and only 3.2 kW(e) is self-consumed. Here 1 kW(e) of high-grade electrical energy is assumed to equal approximately 3 kW of heat energy.

It is important to point out that the per capita energy consumption includes each person's share of the fuel energy consumed by transportation fleets (cars, buses, aviation, etc.), by the manufacture of goods (automobiles, trains, television sets, etc), by agriculture, by food distribution, etc. The oft-quoted figure of 0.6 kW(e) per person (2.4 kW(e) per home) applies only to the average electric grid energy consumed in rural households of North America, and does not include their consumption of petrol energy or their share in energy for the production of goods or foods they use. The latter energies must be added to their 'electric bill' since primary electricity must manufacture all portable synthetic fuels when oil is gone.

Because so much misinformation has been disseminated about nuclear power generation, after the introduction, the book starts out with a review of facts and fables about nuclear energy in Chapter 2. Chapter 3 is perhaps the most important part of the book. It gives the reasons why everyone, including anti-technology environmentalists, should endorse uranium-fueled power generation lest they want to be responsible for globe-warming CO_2 from burning coal and coal-derived synthetic fuels in the next decades. If they abandon nuclear power and do not wish to burn coal either, they must be prepared to eliminate 90% of the world's population after oil runs out. Chapter 3 reviews current energy consumption rates by humans on planet earth and the finite reserves of the world's prime energy sources. The numbers show irrefutably that if we want to reduce global warming and save coal for making plastics and synthetic hydrocarbons, only full exploitation of nuclear energy can save mankind from economic collapse when fossil fuels run out. Wind and solar electric power generation is helpful but only a band-aid. It could never provide the enormous quantities of energy needed to replace all oil-derived fuel with synthetic fuels to sustain our transportation fleets (cars, airplanes, etc). Similarly, saving energy by better home insulation and improving auto engine efficiencies, is useful and should be encouraged. But they cannot change the fact that oil will still run out, as more people on the planet consume it. Chapter 4 reviews propulsion techniques using portable fuels synthesized with the help of primary nuclear electricity. The synfuels should be tested and available by 2020 to start the massive replacement of petrol in order to avoid serious economic upheavals 25 years from now. We *can* survive 2030, but planning and preparations must be started *now*. If replacements of present propulsion technologies and new synfuel productions are not initiated in time, we may see endless wars to control the last remaining oil fields and terrorist activities that make Al Qaeda's atrocities pale by comparison.

Following the first four chapters, back-up material is presented on key features of nuclear reactors, environmental concerns, radiation physics, and security issues, which have been questioned by so many. Factual material for this book was gathered from numerous reports, books, and journal articles written by nuclear professionals worldwide. With the exception of a few always-present contrarians, the vast majority of some 250,000 professional engineers in the nuclear energy field are in agreement with the material presented in this book.

In summary, this book is an admonition that the world must stop vilifying nuclear energy and rapidly expand this unique energy source so we can overcome the pending out-of-oil crisis. Nuclear proliferation concerns and nuclear security issues can be and must be resolved. We don't have any choice if we want to survive oil depletion and stop burning globe-warming coal. Some recent books acknowledging the up-coming energy crisis, project it as unescapable and forecast gloom and doom for our progeny. This book refutes such a scenario and shows a workable solution based on proven technology and scientific facts. Nuclear power should be supported by everyone, in particular by environmentalists. Providence is giving mankind one clear (and clean!) way out of the predicaments it faces when oil is no longer available.

Most derived numbers in the book are scientifically reasonable estimates. But because no one can predict exactly how many people will populate our globe in the next thirty years or how much energy each will consume, and no one knows the precise cost of (future) hardware, there are minor variations of some numbers in different chapters, with which some 'preciseniks' may find fault. They do not invalidate the main conclusions of the book however. Most of this book was written in 2003 and 2004, but at the publisher's request a discussion on Risk Analysis was added as Subchapter 1.2 in 2005. As this book was readied for publication, my attention was called to J.H. Kunstler's recent book titled *The Long Emergency* and Alan E. Waltar's *Radiation and Modern Life; Fulfilling Marie Curie's Dream*. Kunstler expresses many of the same concerns about the pending energy crisis as I do, but offers no solution to overcome this crisis. My prognosis is more optimistic, provided the public listens to 'hands-on' nuclear energy engineers, and ignores the anti-nuclear rhetoric of armchair philosophers who quote fictitious science. Waltar's opus complements this book and shows how important and unavoidable nuclear energy already is in our daily life.

Jeff W. Eerkens
November, 2005

LIST OF BRIEFS

	Future Shock? (Cartoon)	4
1.	Energy and Power Units and Conversion Factors	34
2A.	Annual Energy Resource Consumption in the USA (2000)	35
2B.	Present (2004) World Reserves of Prime Energy Resources	36
3.	Hybrid Electric/ICE Powered Car	58
4.	Flywheel Battery Components	60
5.	Illustration of Internal Combustion Engine Four-Stroke Cycle	62
6.	Typical Fuel Cell Systems	65
7.	Fuel-Cell-Powered Hybrid Car	67
8.	Electricity from Coal – Flowsheet	70
9.	Schematic of Major Nuclear Reactor Components	73
10.	Electric Power Generation by a Steam Turbine	74
11.	Illustration of Reactor Control Rod Operations	75
12.	Typical Layout of a Nuclear Power Plant	77
13.	Nuclear Fission and Breeding Reactions	79
14.	Differences Between a Nuclear Reactor and a Nuclear Bomb	80
15.	Schematic of the Uranium Fuel Cycle	83
16.	Illustration of the Uranium Fuel Cycle	83
17.	Isotope Enrichment Schemes for Gaseous Uranium Hexafluoride	86
18.	Radiowaste Disposal Schemes	89
19.	Long-Term Hazards of Radiowaste Storage in Salt-bed Excavations	91
20.	'Bad' Radioisotopes for Humans and Molecule	99
21.	Make-Up of a Nucleus, Atom, and Molecule	101
22.	Natural Radiation Exposures	106
23.	Diagram of the Chernobyl RBMK-1000 Reactor	115
24.	Chain of Events after Unilateral Moratorium on Nuclear Plant Constructions	122
25.	Required Growth of Nuclear Power to Balance Oil Depletion	135
26.	Future Global Energy pie	136
	No More Oil !!?? (Cartoon)	137

ABSTRACT

The importance of uranium-generated nuclear power is discussed. The review forecasts severe shortages of oil and natural gas in 25 years, and their depletion in 40 years. With a moratorium on burning coal to prevent global warming, it is shown that uranium-produced electricity and heat is essential for large-scale future productions of synfuels (hydrogen, ammonia, hydrazine, and alcohol) made from air, water, and sunshine to replace the portable petro-fuels presently propelling our vast transportation fleets. Generation of 'renewable' solar and wind power is helpful but, unlike nuclear energy, is unable to provide the enormous quantities of prime energy needed for economic manufacture of synfuels to replace fossil fuels. Uranium fission power as a primary energy source is available for 1,500 years to provide all the world's needs for electricity and synfuels at an affordable cost. By manufacturing synthetic fuels from air and water with nuclear heat or electricity, non-portable nuclear energy is in effect converted to portable fuel energy.

The introduction of the book outlines its objectives and reviews four options for combating the upcoming out-of-oil crisis. The second chapter lists common misconceptions about nuclear energy to allay public fears of this all-important energy source. It is followed in a third chapter with a survey of energy consumption rates on planet earth. The fourth chapter is devoted to synfuel production, engines, and fuel-cells that convert the chemical energy of synfuels into locomotion. The last four chapters cover nuclear reactor operations, radiation effects, safety, nuclear weapons proliferation issues, and action items. A university-level education is not required for understanding the book, but exposure to some high-school physics and chemistry will be helpful.

CHAPTER 1

INTRODUCTION

1.1. DECLARATION OF INDEPENDENCE FROM OIL

'What do we do when the well runs dry, my honey? What do we do . . . . . ' This well-known folk-song reflects a very basic problem facing all humans. When the song was written more than a hundred years ago, it referred to the American pioneers who settled on the prairies of the Midwest, and whose water needs depended totally on wells. Without water man cannot survive. Today we can probably assert that without petrol (= gasoline, diesel, etc), most of us would perish. Our water is pumped to our house by pumps run on petrol[1]. Food is brought to our markets by trucks run on petrol. We drive to our jobs and stores in cars run on petrol. Farmers grow crops using machinery run on petrol. We visit family and friends in distant cities traveling on airplanes, ships, or trains, run by petrol. And so on. Clearly our modern world depends totally on refined portable petro-fuels, be they petrol, diesel oil, aviation gasoline, petroleum, natural gas, or other derivatives of fossil fuels. These fuels, which were created by sun-grown plants (solar energy) that accumulated and decayed on the earth surface for millions of years, are being burned up by us in less than two centuries.

Many people tacitly assume that petrol to run our cars, trucks, airplanes, trains, and ships will be available forever. If asked what they would do if there were no more petrol, they often answer 'Our engineers will figure out something', or 'It won't happen during my lifetime so I don't worry about it'. My engineering colleagues who have studied the world's energy requirements and consumptions, are worried. What bothers us most is that there are workable solutions to the problem which are ignored or obstructed by a number of governments in the world because of misunderstandings and ignorance. Some government politicians have been pursuaded by special-interest groups to subsidize costly 'alternative energy' projects without consulting senior experienced energy providers and technologists

---

[1] The British 'petrol', called 'gasoline' in the USA, 'benzin' in Germany, and by other names, will be used from here on for fuel extracted from refined crude oil. Diesel fuel is also assumed to be part of petrol supplies.

who have handled power generation for decades. They are funding studies of alternative energy programs currently in vogue such as wind farms and solar collectors, corn-derived ethanol, etc. This lulls the public into believing that everything is under control while it is not. These 'renewable' power-generation schemes can only be supplemental as discussed later. Compared to nuclear, they are economically much inferior to provide the large amounts of electric energy needed for the manufacture of synthetic fuels to replace present oil-derived fuels consumed by the world's transportation fleets. New nuclear energy programs must be funded *now* to prevent a serious energy crisis in the near-future.

This book is an alert and warning to heed the call for an expansion of nuclear electric power to stave off future shortages of fuels for the propulsion of cars, trucks, trains, aircraft, and ships. By the most optimistic estimates, at present rates reasonably accessible oil and gas reserves will be depleted in 40 years, if consumption continues to increase at current levels. Oil retrieval rates have already been declining since the mid-1990's. Unless preventive measures are taken immediately, steadily increasing oil shortages will reach a crescendo by 2030, triggering a total collapse of our present oil-dependent way-of-life. New propulsion systems must be developed which can use synthetic fuels (hydrogen, ammonia, hydrazine, etc) obtainable from air and water via nuclear (or coal) heat or electricity. Uranium and thorium can provide the prime energy needed for massive production of these synfuels for more than 1500 years, and coal for approximately 150 years. However coal should be preserved for the manufacture of plastics and other organics-based materials when oil is gone. In addition, coal combustion promotes global warming, and its use in generating electricity should be replaced with non-air-polluting nuclear power plants. Unless one wants to promote a 90% reduction of the present world population after 2030 via famine and wars, the only way to prevent a serious economic crisis twenty-five years from now, is to start the immediate expansion of uranium-consuming and -breeding electric power plants. Such undertakings require a lot of preparation, and unless misinformed lobbyists and politicians reconsider their opposition to nuclear power, humanity will be heading straight into a storm without fuel.

There are those who believe that nuclear reactors are too closely tied to nuclear weapons, and since they wish to ban nuclear weapons, they automatically oppose any expansion of nuclear power. By presenting them with the facts, it is hoped that most of them will change their minds. Some have convinced themselves that if government would only help develop conversion of garbage into fuel, and forced everyone to cover their house with solar panels and/or a wind turbine on their roof, all future energy problems will be solved. However this is simply not realistic. A similar 'small-is-beautiful' notion was once proposed and carried out in the 1960's by the communist regime in China, which had every community build a furnace to melt down scrap iron to meet national needs for steel. The whole scheme quickly fizzled because it was not cost effective.

In the present energy case, one forgets that for manufacturing autos, aircraft, bridges, houses, etc., and for transport of goods and people by cars, trucks, trains, ships, airplanes, enormous quantities of electricity and petrol are consumed. This energy load must be shared by every man, woman and child. Windfarms are great for low-power applications in wind-blown regions of the globe, but they cannot practically and economically solve the global fuel shortages we face twenty-five years from now. Solar-cells, wind-mills, and energy-conservation measures for homes and kitchens are useful, but can only provide a minor assist to reduce global oil consumption; they may extend the 'out-of-oil' date by a few years. Solar and wind energy industries have been generously subsidized for more than thirty years. Their products have matured and found many markets for small-quantity electricity needs (remote homes and facilities, traffic signals, sail-boats, etc). But to be considered as electricity providers to feed heavy industries and to manufacture oil-replacing synfuels for the transportation fleets of the world, they must be able to withstand the test of a rigorous engineering evaluation. Subchapters 1.2 and 3.3 shows that the 'renewables' fail that test when competing with non-air-polluting nuclear power. Uranium-generated electric power which can remedy the upcoming out-of-oil crisis, is a *long-term 'green' energy fix*. Solar and wind power qualify only as *expensive 'green' energy aids*.

Many non-technical people seem not to appreciate *scale-ups (magnitudes)* and supply *rates*, two concepts very familiar to engineers. Clearly a 100,000 gallon (378,500 liter) storage tank filled with corn- or biomass-derived alcohol fuel, could never feed a million cars in a big city if each car burns 2 gallons (7.57 lit) per day on average, and corn-growers can only provide enough alcohol on a continuing basis to refill the tank once a day. It could take care of 50,000 cars but not 1,000,000. Yet promoters, aided by the media and some celebrities, insist that production of alcohol fuel derived from corn, switch-grass, or other biomass source, is the solution to our pending out-of-oil crisis; nuclear is not needed. They don't realize that without (nuclear) electricity, one needs twice the entire land surface of the USA to produce enough alcohol for its fleets of cars, trucks, and airplanes to replace petrol consumption (Sections 1.2.2 and 3.3). The oil we are burning up in two centuries represents biomass from decayed plants that were originally energized by the sun over a period of many hundred millions of years. Since the solar flux at the earth surface has not changed much since that time, and energy produced must equal energy consumed, one cannot expect that one is able to match our present oil consumption rates solely by growing new biomass a million times faster, even with more arable land and modern agricultural sophistications. Only with a major energy input of primary nuclear electricity to run farm equipment and distill out alcohols, can one replace US petrol consumption if one-third of all land in the USA is planted with bio-fuel crops.

The energy concerns expressed in this book were focussed on the USA for which statistical data were readily available to the author. However it is clear they apply equally to all nations in the world. Many parameters (e.g. needed replacement fuels) can be scaled in proportion to the population in each country and their degree of prosperity. Governments should focus on exploring realistic solutions to prevent the large-scale fuel shortages that will develop in the not-so-distant future when

4   CHAPTER 1

FUTURE SHOCK?

oil runs out. They must recognize the critical problems correctly and assist with solutions that may require billions of dollars which only they can lend or allocate. They are often misled by opinionated anti-nuclear political activists who have a 'kill the messenger' mentality when confronted with unpleasant facts. Former USA House Speaker Sam Rayburn once said: 'Any jackass can kick down a barn but only a skilled carpenter can build one'. It is time we kick out the jackasses and let the carpenters go at it!

1.2. NEEDED ACTIONS AND RISKS TO OVERCOME THE PENDING NO-OIL CRISIS

This book should convince most technologists that there is an urgent need for more nuclear power to overcome the pending energy crisis. However environmentalists, public policy makers, and financiers like to see a risk assessment of the nuclear option compared to other possible solutions [Refs 50, 51, 52]. Such stakeholders want to be convinced that an expansion of nuclear power plants can be done safely and economically before they will give it their support. They are apprehensive because a multitude of fear-instilling misrepresentations have been circulated in the media about nuclear power (see Chapter 2). Many of these stakeholders favor development of more solar, wind, and biomass energy at the exclusion of nuclear energy without considering the scale, cost, and feasibility of these so-called 'renewables' to replace the vast quantities of portable fuels presently extracted from the oil fields.

We define the risk of an undertaking as the probability of encountering negative effects involving: (a) hazards to human health, safety, welfare, environment; (b) technical feasibility, operability, the ability to deliver; and (c) financial feasibility. We further assume that all stakeholders believe in a democratic world that favors the well-being, health, happiness, and freedom of all citizens on our planet. That is, we exclude consideration of dictatorships and societies that would not hesitate to eliminate a substantial portion of the world's population to balance energy supplies and demands for the benefit of a few.

While items (b) and (c) are reasonably quantifiable, (a) is more difficult to measure since it involves intangible human fears and perceptions. As discussed in Reference 50, most people give more weight to one large accident that kills 3000 humans than 3000 single deadly traffic accidents, even though the same total number of fatalities occur. Also a loss of $100 in the stock market often hurts more than the satisfaction of gaining $100. That is, values (enormities) in sociometry are often not linearly related to some calamity parameter (a human death or a monetary loss). This skewed nature of human perception gives rise to numerous disagreements between stakeholders, and sometimes leads to the selection of non-optimal solutions when alternatives are considered. To minimize controversies, it is necessary to carefully define objectives and risk factors, and to insure that the concerns about nuclear power of all stakeholders are recognized. Such concerns are repeatedly addressed throughout the book.

In the present case we want to find the best means of overcoming the pending no-oil energy crisis, and we shall assess the risks associated with (1) the 'nuclear millennium option', (2) the 'renewables-only option', (3) the 'century-long coal-burning option', or (4) the 'do-nothing option'. The postulated scenarios of these four options are as follows:

1. The Nuclear Millennium Option. In this option we propose to use heat or electricity from nuclear power plants to manufacture synthetic fuels (synfuels) to replace present portable fuels derived from oil. Also, all coal-burning power plants are replaced with nuclear plants, that is all electricity is derived from uranium fission. This program requires a ten-fold expansion of all existing nuclear power plants and the introduction of breeder reactors [Ref. 32]. Concurrent development of new automotive engines (fuel cells or combustion engines) which consume the new synfuels is also part of this scenario. Further, all railroads are electrified (i.e. diesel replaced by electric motors) to minimize the dependence on synfuels.
2. The Renewables-Only Option. In this approach, massive deployment of solar panels and windmills are proposed to supply all prime energy for electricity and for making synfuels. Electric energy obtained from solar and wind power stations is used to manufacture synfuels as in (1). Also agricultural crops and forestry (corn, soybeans, wood, etc) are to provide bio-fuels which are mostly alcohols and heavy oils.
3. Coal-Burning Option. The global-warming threat is ignored, and heat or electricity obtained from additional coal-fired plants is used to manufacture

synfuels as in (1). Option (3) is identical to option (1) except coal replaces uranium. Option (3)with manufacture of synfuels to replace fossil fuels, is only sustainable for one or two centuries compared to one or two millennia if uranium resources are mined.
4. The Do-Nothing Option. As the label implies, under this option nothing is done. A massive world-wide recession ensues as oil runs out. The world will return to primitive living conditions that existed before the 19th century, except there are ten times more mouths to feed [Ref. 53].

In what follows we briefly examine risk categories (a), (b), and (c) for options (1), (2), (3), and (4).

1.2.1 Risks in the Nuclear Millennium Option

1.2.1.1 Safety Considerations and Biosphere Effects

The most debated safety issues raised by environmental groups regarding nuclear power programs are: (i) Disposal of nuclear wastes; (ii) Radiation effects on people due to reactor accidents caused by unintentional or intentional (terrorist inflicted) malfunctions; (iii) intentional diversions and modifications of nuclear fuel to make nuclear weapons.

In Chapter 5 we show that the disposal of radioactive waste can be carried out safely. Even in the event that an accident occurs during transport of spent nuclear fuel elements from a reactor site to an underground storage facility, the exposure of the public to radiation is virtually nihil. From transportation statistics and collision tests with armored nuclear caskets it is estimated that one out of every 100,000 radioactive material transports might experience an accident in which the transport casket is penetrated via a crack or terrorist bullet hole. The risk that someone in the public is subsequently exposed to harmful radiation due to such a breach is estimated to be less than 10^{-4} so the overall probability of a harmful radiation exposure due to the movement of radioactive materials is less than 10^{-9} per transport.

Concerns about diffusion of radioactive chemicals through soil from a nuclear depository to aquifers tapped for drinking water have been extensively studied [Ref. 33 and Brief 19]. The probability that a nuclear storage casket will leak and be exposed to leaching ground water during 100 years of underground storage is on the order of 10^{-3} while the probability that a person will get cancer after drinking radioactive water from soil-penetrating chemicals that migrated from a nuclear storage site to an aquifer thirty miles away is on the order of 10^{-4} in a period of 100 years (Brief 19). Thus the overall risk that a person may get cancer from drinking contaminated water near a nuclear storage site is on the order of 10^{-7} in 100 years or 10^{-9} per year. This compares with a chance of 10^{-4} per day for an automobilist to have a collision and a probability of 10^{-7} per flight to be in a plane-crash for air travelers.

The probability of a death in the public outside a nuclear power plant due to a reactor accident is discussed in Chapter 6. The only public deaths due to a nuclear power plant accident occurred in 1986 at Chernobyl in the Ukraine. The total death toll including thyroid irradiation victims who (unwarned) drank contaminated milk from cows that had grazed on grass covered with particulate fall-out, came to 50 people. Predictions of additional future deaths due to latent cancers induced by radioactive fall-out are without foundation and lack credibility (Chapter 6). After 10^7 MW-years of worldwide nuclear power that was generated over the past thirty years, the observed risk of causing a human fatality in the public due to a reactor meltdown is thus 5×10^{-6} per MW per year, based on the one accident at Chernobyl. For a typical 1000 MW reactor the worlwide risk of causing a human death would then be 5×10^{-3} fatalities per year (one death in 200 years). In comparison, the worldwide risk for underground mining is more than one fatality per year per mine.

In the USA, which presently generates about 10^5 MW with 100 reactors, there was one reactor meltdown in 1979 at Three-Mile-Island (TMI). In this maximum credible accident, no one in the public was over-exposed to radiation thanks to the reactor's heavy containment vessel that confined all radioactive debris. Thus for the USA alone, the probability of causing a fatality due to a nuclear reactor meltdown is 0 per year per MW after 40 years of operations. After the TMI and Chernobyl accidents, many new sophisticated accident prevention techniques and public safety measures were introduced at all nuclear power plants in the world. Russia has (and nearly all other nations have) also adopted most new reactor safety features now mandated in the USA and the EU. This makes another Chernobyl-like accident virtually impossible (see Chapters 5 and 6). For a modern water-moderated 1000 MW(e) power reactor with a properly designed containment vessel (absent at Chernobyl), the probability of public exposure to particulate fall-out radiation after a reactor core meltdown or destruction (by earthquakes, planecrash strike, or a suicide terrorist), is estimated to be less than 10^{-6} per year. Assuming all future reactors are built under the same rigorous regulations imposed today, this translates to a risk of 9×10^{-3} public radiation exposures per year (1 in 110 years) for all 9000 reactors expected to be in operation worldwide by 2050.

Issue (iii) under (a), that is the diversion of nuclear reactor fuel to make nuclear weapons by increasing isotope enrichments or extracting and purifying neutron-bred plutonium, is discussed in Chapter 7. As explained in Chapter 7, the probability for a terrorist group to make a nuclear weapon is extremely remote unless they are aided by a sovereign nation. The theft of a completely functional nuclear weapon from a heavily guarded military facility is also very unlikely as discussed in Subchapter 7.2. Only an independent country could design, build, and test a nuclear weapon at great expense (over $ 1 billion) if its government wants to do so. The chance of such an event is independent of the operating risks associated with building nuclear reactors for manufacturing synfuels. With a strong internationally monitored Non-Proliferation-Treaty (NPT) program (Subchapter 7.3), it appears that the probability of nuclear weapons proliferation can be held to a low level if all countries sign up for the NPT and abide by it. There is no reasonable logic to add this risk in the assessment of the nuclear millennium option (1). Avoidance of building future nuclear power plants under option (1) will not prevent any determined nation from developing nuclear weapons. The know-how to do so is readily available in the world literature. Today, with a few billion dollars any sovereign nation can avail itself of a nuclear weapon in five years.

Regarding effects on the biosphere, nuclear power plants produce no air pollution or globe-warming gases like coal-fired power plants do. All 'burnt' uranium products remain encapsulated in solid fuel elements which can be safely disposed of as discussed in Subchapter 5.2. In summary, type (a) risks involved in developing more nuclear power plants are several magnitudes less than many commonly accepted industrial endeavours that keep us alive on planet earth. The fact that nuclear power generation produces absolutely no air pollution or global warming is a strong plus. On the negative side, concern about nuclear weapons proliferation has added a public fear factor and an irrational opposition for 'things nuclear'. But as discussed in Chapter 7, weapons proliferation is a separate issue independent of implementing option (1). While the same uranium is used as raw material (but isotope-enriched to very different levels), a nuclear power plant is entirely different from a nuclear weapon, and one cannot be converted into the other as explained in Section 5.1.3. It would be as senseless to halt nuclear power plant development as it is to stop making jet aircraft or bulldozers because the latter could be converted into military fighter planes and tanks, or because their materials of construction can be diverted to making bombs.

When oil fields run dry, portable synfuels must replace today's highly developed petrol fuels. They must be readied for mass production concurrently with an expansion of nuclear power plants. Simultaneously, all petrol-burning engines must be replaced by new engines powered by the new portable synfuels. Since the massive quantities of carbon dioxide gas exhausted from present engines appear to cause excessive global warming, exploration of new synfuels has focussed on those that do not add any new carbon dioxide to the atmosphere. These include hydrogen, ammonia, and hydrazine synthesized from water and nitrogen from the air, whose exhausts are again non-polluting water and nitrogen after a reaction with atmospheric oxygen in the new engines. In addition, biofuels such as alcohols obtained from plants with the aid of electricity, which consume carbon dioxide from the air during plant growth and return it to the air when oxidized, are considered acceptable synfuels. Two types of engines can be powered by the new synfuels, namely internal combustion engines (ICEs) and electrochemical fuel cell engines (FCEs). Chapter 4 gives more details of possible future synfuels and engines. Under option (1), the prime energy needed for fuel synthesis is provided by nuclear electricity or heat so that in effect non-portable nuclear energy is transformed into portable fuel energy. Note that suitable synfuels are recycled through the biosphere's enormous reservoirs of water, air, and carbon dioxide without depletion of feed material.

It is estimated that to synthesize hydrogen or extract bio-alcohol with nuclear electricity or heat, would cost 1.5 to 2.5 times more per Joule or BTU than refining petrol from crude oil. Then instead of today's price of $0.50 per liter or $2 per gallon (including tax), the new synfuel price would be about $1 per liter ($4 per gallon), or $ 64 for traveling a typical 600 km (373 miles) on a tankful of synfuel, assuming government taxation rates remained the same. The cost of ammonia or hydrazine synfuel would be slightly higher at 2 to 2.5 times that of petrol per combustible Joule, since the main feed material for making the latter synfuels is pure hydrogen. Aided with nuclear energy, one might consider retrieving carbon-dioxide from the atmosphere directly, reconstituting it into hydrocarbons like the trees and plants do naturally. However there is only 0.03% carbon-dioxide gas in the atmosphere. Reforming it into hydrocarbons takes enormous amounts of energy and would cost twenty times more than making bio-alcohol.

As discussed in Chapter 4, one major research area requiring more research and development is the compression or bladder-storage of gaseous hydrogen to acceptable volumes for automobile travel. Also the safe distribution and handling of bio-alcohol and hydrogen in enormous quantities for public use demands more studies. Hydrogen effuses through many plastics and embrittles a number of metals, so special gas-tight pipes would be needed. For hydrazine, stabilizing additives and special storage tanks have been proposed to minimize possible unintended ignitions or explosions in the presence of atmospheric oxygen. Many experimental programs are under way in industry, universities, and government labs to advance the state-of-the-art of mass-handling the synfuels and to develop durable FCEs and ICEs that use them. Actually the risks of distributing and storing flammable petrol are not much different from those of bio-alcohol, ammonia and hydrazine, but the latter have been less explored. Flammabilities of alcohol, ammonia, hydrazine, and hydrogen in air, while not quite the same, are comparable to that of petrol.

Unless the problems with hydrogen can be solved, bio-alcohol and ammonia appear presently the most acceptable synfuels from the aspect of safety and handling. They can replace petrol in internal combustion engines or in the case of ammonia, feed fuel-cells upon catalytic decomposition to hydrogen and nitrogen on a proton exchange membrane (PEM). Under modest pressure (about 20 atm), ammonia is a liquid and has been used extensively as a fertilizer in agriculture for many decades. Its handling and distribution is well developed and similar to that of petrol. The handling of liquid bio-alcohol should likewise present few probems. Hydrazine is somewhat unstable and decomposes if heated. It is a liquid at room temperature and has been used for rockets. It might fuel future aircraft if the addition of a suitable stabilizing agent for safe storage and handling can be found. Substantial amounts of hydrazine and hydrogen have been used safely in a number of aerospace programs. Clearly handling and distribution techniques of the new synfuels are not new but more investigations are needed (mostly hydrogen and hydrazine) for applications involving the public. In conclusion, the main social risk in carrying out option (1) may prove to be the safe handling and massive distribution of alcohol, ammonia, hydrazine, and hydrogen rather than the nuclear component.

1.2.1.2 Technical Feasibility, Operability, and Ability to Deliver

The technical feasibility part of option (1) faces few risks since nuclear power plants are well developed. They have excellent safety records (except for Chernobyl) during more than 40 years of operation. However to fully achieve more than a millennium of uranium-provided energy, current 'burners' must be modified into 'breeders' in the next twenty years [see Chapter 5 and Reference 32]. Burners only consume the less abundant (0.7%) U-235 isotope of uranium, while breeders or 'fast-neutron reactors' utilize almost all of the uranium, thereby extending the resource availibility period by more than a factor of 60 (from 30 to 1800 years !). Quite a number of experimental breeder reactors have been built and tested in the last forty years and performed as expected. But because they are slightly more expensive and their spent fuel must be reprocessed, the nuclear power industry has built mostly burners so far. Currently only France, Russia, and Japan each have one breeder power reactor in operation and have dedicated programs to replace burners in their country with breeders in the coming decades. Besides the complete utilization of uranium fuel, the advantage of breeder reactors is that final quantities of nuclear waste material are much lower than for burners. National nuclear waste repositories like Yucca mountain can therefore be used for hundreds of years before they are filled, instead of only a few decades under a burner regime. Before new nuclear breeder reactors can be built in great numbers, tests are needed to select a few optimized models for mass deployment. In summary, the technology for building and operating breeder reactors is in place but needs more encouragement by governmant planners and regulators.

Another benefit offered by nuclear power plants besides being non-air-polluters is that in generating electricity, two-thirds of the nuclear-produced energy is low-temperature heat which could be made available for some other application. Presently this heat is dumped into river water or a cooling tower and is thus essentially wasted. In coastal areas where there are fresh water shortages (e.g. California, Arabia), this heat can be profitably used to desalinize seawater. Several dual-purpose nuclear power plants have been designed and investigated in the last forty years, but so far only Russia is operating one. Only when water shortages become severe will there probably be further action to consider this side benefit. Irrational public fear and ignorance has so far hindered this application of nuclear energy in the USA. Besides supplying heat to evaporate seawater, waste heat from nuclear power plants can provide central heating for clusters of buildings in cities with very cold climates such as Ulaanbaatar in Mongolia.

Synthetically manufactured portable fuels and automotive engines powered by them, involve non-nuclear technologies that have been researched and developed for over a century. However no synfuel has been mass-produced on the same scale as the present extraction of petrol fuels from crude oil. Although not handled on the same scale as petrol, large quantities of hydrogen, ammonia, alcohol, and hydrazine have been produced by the chemical industry for agriculture (ammonia), the space program, semiconductor manufacture, and scientific research. A hundred-fold increase in the production of these synfuels should pose no basic technical problems except for possible plant construction delays of the energy-providing breeder reactors that will help produce them. This is because of the enormous increase in labor and materials needed for construction of a thousand new breeder reactors to meet a 2030 deadline (Chap 3).

Regarding development of new internal combustion (IC) and fuel-cell (FC) engines powered with hydrogen-based portable synfuels, there are no fundamental show-stoppers to mass-produce them by the year 2030. Experimental internal combustion engines using ammonia, hydrazine, and alcohol in place of petrol have been built. Although water (H_2O) and nitrogen (N_2) or carbon-dioxide (CO_2) are the main exhausts to the biosphere, because of the high temperatures in ICEs, toxic nitric oxide (NO_x) gases are formed as byproducts which must be catalytically converted and eliminated. Further refinements and tests are thus needed before safe synfuel-burning ICEs can be mass-marketed. FCEs with ammonia or hydrazine feeds which are catalytically decomposed to provide pure hydrogen on PEM's have also been demonstrated. Such FCEs have no NO_x exhaust problems. Unless gaseous hydrogen storage problems can be solved, compressed (20 atm) liquid ammonia synfuel for FCEs and ammonia or alcohol for ICEs appear presently most acceptable for empowering future mass transportation.

While the technical feasibility for implementing option (1) seems assured since needed technologies have been demonstrated, a large challenge will be to field one-thousand new breeder reactors in the USA (nine thousand for the world) by the year 2030, when serious oil shortages will develop and oil-replacing synfuel production must be in place. The hundred reactors presently operating in the USA (four hundred thirty worldwide) will be unable to provide the needed expansion of prime energy to implement option (1). It means that the USA must build on average some forty new breeder reactors each year for the next 25 years which will put a big strain on the availability of skilled labor and materials. The most practical approach initially would be to double and triple the number of nuclear reactors at existing reactor sites, since these sites have already been cleared for reactor operations. New sites must ultimately be developed however to accomodate the new nuclear synfuel age. The overall effort would be comparable to WW-II when the USA produced hundreds of military aircraft, ships, and tanks weekly. We have a war for survival on our hands which must be won to avoid an economic disaster and collapse of our civilization after 2030. Leaders like Admiral Hyman Rickover and General Leslie Groves (Section 5.2.1) are needed to help us get through the next two decades. If nothing is done (Section 1.2.4) there will be no more aviation fuel to fly our airplanes, no more diesel to move our food and freight by trucks, trains, or ships, and no more petrol for cars and buses. The engines in all these vehicles will be useless and must be replaced with new FCEs and ICEs. Due to public opposition to nuclear power, today the number of companies in the USA capable of designing and building nuclear breeder reactors has dwindled to two, from an original eight or so in the 1960's (the goal was then to build 300 power reactors by 2000). It is likely that most of the new reactors will be built by French, Japanese, Russian, Chinese, and Indian firms who have overtaken nuclear power plant design and construction, a field once dominated by the USA.

1.2.1.3 Financial Feasibility

Presently (2005), the cost of building a 1200 MW(e) nuclear power plant is approximately $1.8 billion. To build one thousand new reactors by 2030 would then cost a total of $1.8 trillion in today's dollars, or $72 billion per year for the next 25 years. With the USA's annual GDP on the order of $10 trillion, this level of development is sustainable financially, particularly if economic survival is at stake. As mentioned however, to build forty reactors per year in the USA (three hundred sixty worldwide) for the next 25 years, will require an enormous build-up of skilled labor and materials. Shortages are almost certain to develop initially which will impede progress. Because of this, the nuclear build-up program may have to be stretched out to 50 instead of 25 years, and global warming may have to be tolerated for one or two decades longer, to have coal-burning power plants provide energy for synfuel manufacture (Section 1.2.3) until they can be replaced by non-air-polluting breeder reactors. The risk of a reactor meltdown, estimated to be 1×10^{-3} per year for all 1000 reactors in the USA (10^{-6} per reactor), requires an insurance policy costing not more than $1 billion per year ($1 million per reactor per year) to cover public liability and physical plant damage.

1.2.2 Risks of the Renewables-Only Option

1.2.2.1 Safety Considerations and Biosphere Effects

The most quoted 'renewable' sources of energy are solar, wind, and biomass. Solar panels have been extensively developed since the beginning of the space program in the 1960's, while modern wind-driven turbo-generators have seen an enormous growth in the last two decades. These energy sources have gained a lot of popularity and are useful for providing electricity in small-scale applications. However they are limited in capacity and cannot be deployed economically on the large scale needed for replacing the portable fossil fuels that presently propel our vast transportation fleets and heavy industries (see below and Subchapter 3.3). In regard to type (a) risks, at first sight, solar panels seem not to pose any environmental risks. However if they were to be produced in very large numbers, one must consider the toxic hazards associated with the use of arsenic, selenium, indium, gallium, and siloxanes in their manufacture. If one considers generating 1 TW (10^{12} Watt) of electricity with solar cells to satisfy the USA's oil-replacement requirements, one would need an inventory on the order of 1,000,000 tons of these special chemicals, assuming it takes 1 gram per solar Watt. Since solar cells deteriorate (due to sand erosion, bird droppings, etc) they must be replaced and recycled every ten to twenty years. This means one must handle 50,000 tons of poisonous arsenic, gallium, etc per year which presently exceeds the mining capacity of these chemicals. This compares with about 5000 tons of uranium per year to fuel one thousand breeder reactors which together generate 1 TW. In addition one would need approximately 25 million acres (100,000 km^2) of sunny desert land to extract 1 TW of year-averaged solar power.

To generate a year-averaged 1 TW of wind power requires 1,250,000 windmills of 2 MW(e) assuming an average of 20% wind availability. The problem with windmills is that they kill thousands of birds and spoil many landscapes if deployed over 30 million acres (120,000 km^2) of windy land. Environmentalists in Massachusetts and in the Netherlands have protested the further expansion of windmills because of this. Regarding the use of corn-extracted alcohol and biomass to fuel our transportation fleets one finds that unaided by electricity, the USA would need 3 billion acres of arable land to provide 1 TW of net biomass energy. This is more than the entire surface area of the USA of 2.24 billion acres (one finds the same result for the rest of the world; see Subchapter 3.3). However if nuclear energy is used to help cultivate and harvest corn or sugarcane and to extract its alcohols, only 0.6 billion acres are needed which is doable. Since carbon-dioxide from the air is converted into bio-alcohol fuel and later returned to the atmosphere upon fuel combustion, there is no net globe-warming carbon-dioxide addition to the biosphere.

Risks associated with developing facilities to manufacture synfuels with prime electric energy from solar or wind power complexes are similar to those in option (1). Like in option (1), the storage and distribution of synfuels for public use will need close examination to insure public safety.

1.2.2.2   Technical Feasibility, Operability, and Ability to Deliver

As already alluded to, while technologically developed, solar and wind power require enormous land areas. Although empty inexpensive desert lands could be employed, a large network of energy storage units (batteries) and distribution/collection power lines would be needed, which adds to capital and maintenance costs when compared to a single coal or nuclear power plant housed in one building. If 1 TW is needed for oil replacement, solar alone would require more than 1% of all of US territory and wind power would need 1.5% of all its land mass. If one grows bio-fuel-producing plants with exclusion of nuclear energy inputs, one could never replace all present fossil fuels since it requires more than all arable land in the USA. Only with the aid of uranium-generated electricity could bio-fuels replace petro-fuels, provided one-third of all lands in the USA are made available for cultivation of bio-fuel-producing plants.

The need for a million tons of special chemical materials to manufacture and install solar cells for the generation of 1 TW of solar power, and the annual reprocessing or replenishment of 50,000 to 100,000 tons to maintain this level, exceeds presently available mineral resources from which they are extracted. Such large quantities may well be physically unobtainable.

1.2.2.3   Financial Feasibility

As discussed in Subchapter 3.3, the cost of producing 1 TW of power from solar energy stations is calculated to require an investment of about $ 9.1 trillion, while wind power would need $7.2 trillion compared to $ 1.8 trillion for nuclear power. With a US gross domestic product (GDP) of $ 10 trillion, the commitment to a renewables-only program for replacing oil, would strain the financial community. Although fuel costs are 'free' for wind and solar power, maintenance costs are very high because of the large expanse and number of solar panels and wind-turbines. Compared to coal and nuclear power generation, one finds solar and wind power to be at least four to five times more expensive at the 1 TW level. This applies to both capital and operating costs. (For nuclear power plants, fuel costs are less than 30% of operating costs.) Costs for insurance against natural disasters such as hurricanes, sand-storms, and earthquakes which could damage vast areas of solar or wind power complexes, must be added to overall operating expenses of course. It will likely be two to three times higher than insurance for the nuclear option (1), because of the large land areas covered (higher exposure to natural forces).

1.2.3 The Coal-Burning Option

1.2.3.1 Safety Considerations and Biosphere Effects

Conversion of coal energy into mechanical and electric energy via a steam cycle, has been utilized for centuries. Coal-burning power plants can be substituted for uranium-consuming nuclear plants (and vice versa) to supply the energy needed for manufacturing synfuels that replace fossil fuels obtained from oil. However coal reserves would be depleted in 100 to 200 years compared to known uranium reserves that would last for 1000 to 2000 years in a breeder reactor economy. In addition and even more important, coal-burning produces carbon dioxide gas which has been found to cause irreversible warming of the earth atmosphere due to the green-house effect. If coal were to be converted into petrol by the SASOL process (Section 4.1.1), the globe-warming problem is not solved since the combustion of carbon-carrying petrols also creates carbon-dioxide gas in automobile exhausts. It has been proposed to capture and sequester carbon dioxide emissions from coal-burning power plants and automobiles instead of releasing it in the atmosphere, but this requires extra energy and will greatly increase the cost of using these fuels.

Besides producing globe-warming gaseous carbon dioxide, coal-burning plants also inject air-polluting mercury, uranium, and other undesirables entrained in particulate matter into the atmosphere. Scrubbers recently installed on coal-burning plants to remove particulate emissions, have lessened air pollution but considerably increased the cost of using coal energy. Carbon which makes up coal, is an essential element in all compounds widely used in plastics, paints, lotions, pharmaceuticals, etc. This raw material should be preserved for future generations, and not burnt and dispersed as carbon dioxide through our biosphere.

It is difficult to determine the risk factor for burning coal (and petrol). At international conferences on global warming, it is claimed that millions of people will be displaced and die, when climate changes due to global warming cause flooding of coastal areas and islands. Also it is forecasted that global warming will increase hurricane activities. Finally, mercury inhalations from polluted air cause additional casualties.

Risks involved with manufacturing, storing, and distributing synfuels were discussed in Section 1.2.1 and are the same for options (1) and (3).

1.2.3.2 *Technical Feasibility, Operability, and Ability to Deliver*

Coal has been a major prime energy source for more than two centuries, and technologies using its heat of combustion are well developed. Coal is more abundant than the petroleum fuels found in the earth crust. As shown in Chapter 3, it may be available for up two centuries if used as the only prime energy source for synthesizing portable hydrogen-based fuels to replace fossil fuels. If option (3) is chosen, a four-fold expansion of coal-burning power plants would be needed over the present number of power plants, together with a four-fold increase in coal mining operations and coal transportation by rail.

Carbon dioxide gas has been pumped into depleted underground oil reservoirs in some experimental programs in Norway, but it is not certain if such a sequestration technique can be carried out succesfully if coal energy is consumed on a tera-watt scale.

1.2.3.3 *Financial Feasibility*

The total cost of a four-fold expansion of present coal-fired power plants in the USA would be approximately $ 1 trillion for seven hundred and fifty additional 1200 MW(e) coal-fired power plants at $ 1.3 billion each. Costs for a four-fold expansion of current rail transport of coal are uncertain but may add $ 80 billion overall. These costs are affordable if they avoid an economic meltdown.

1.2.4 The Do-Nothing Option

1.2.4.1 *Safety Considerations and Biosphere Effects*

A worldwide economic recession will result under option (4), after oil sources dry up. Famine and warfare will break out, causing the deaths of many urban dwellers and combatants. Warfare results because of a desire to conquer and control the last remaining oil fields which will be defended by the military. The world's population will decrease severely in one generation due to the inability to transport adequate amounts of food, water, and goods (no petrol), resulting in starvation and disease (as experienced on a small scale in New Orleans after hurricane Katrina). Some environmentalists may applaud a decrease in the world population, but this sentiment will not be shared by those who wish to live and survive.

1.2.4.2 *Technical Feasibility, Operability, and Ability to Deliver*

Clearly doing nothing is quite feasible and often the course taken by a laissez-faire public and undecisive bureaucracy, until a calamity strikes [Ref. 53].

1.2.4.3 *Financial Feasibilty*

Since no funds have to be expended under option (4), this path of no resistance is certainly feasible.

1.2.5 Conclusions based on the Risk Assessment of Options (1), (2), (3), and (4)

In a democracy, option (4) would in principle be rejected outright, since too many people would perish. This leaves us with options (1), (2), and (3), or a mixture of them. In all of these three options, a program for extensive synfuel manufacturing is included which is to be developed concurrently with an expansion of prime energy providers, whether uranium, renewables (solar and wind), or coal. The technology of manufacturing synfuels is essentially identical and can be evaluated independently of possible prime energy suppliers. In Chapter 4 it is shown that electrical and mechanical storage batteries are impractical for long-haul applications as alternates for inducing locomotion. Only synfuels produced with the aid of nuclear electricity from air and water (and sunshine for bio-alcohol) are adequate substitutes for current petrols.

Several different processes are available to synthesize hydrogen, ammonia, and hydrazine from air and water, some of which are discussed in Chapter 4. While new nuclear power plants are being designed and constructed to supply needed synthesis energy, the most efficient chemical synthesis schemes must be selected via pilot plant experiments. This can be done with non-nuclear heat or electricity. Simultaneously, tests should be conducted to explore concepts for safe handling and compact storage of the new synfuels for applications in the public automotive world. This latter task is very important and should be completed within the next 15 years. Finally in another concurrent 15-year development program, tests must be carried out with new ICE and FCE designs fueled with ammonia, hydrazine, hydrogen, and/or alcohol. A determination must then be made made which engine would be most suitable for public automobiles, which for buses, which for trains, which for ships, and which for aircraft. The three programs of synfuel production, synfuel handling, and new engine development must be integrated and timed so that they can be ready for mass production when the first synfuel production plants come on stream.

In addition to portable synfuel production, it would be prudent for the USA to (re-)electrify transport systems (trams, trolleys, buses, and trains) nationwide as much as possible, like is done in many parts of Europe. Concepts to provide cars with electric power via overhead electric lines on major freeways as is done for electric trams and buses, might also be implemented. When the cost of petrol sky-rockets in the 2020's due to oil shortages, 'hybrid' autos (Chap 4) might be powered by petrol-burning combustion engines when away from overhead electric lines, but use electric motors for locomotion when overhead electric lines are available on freeways. Whether automobile travel can be (partially) electrified or not, it is clear that additional nuclear power plants must be built to replace the energy of substituted petrol.

Next we investigate the prime-energy-providing portions of options (1), (2), and (3). For generation of 1 TW in the USA, option (3), coal-burning, is the least expensive at $ 1 trillion, compared to $ 1.8 trillion for nuclear, $ 7.2 trillion for wind, and $ 9 trillion for solar energy. However option (3) will be rejected by environmentalists who believe global warming and air pollution is a serious threat that must be avoided. Also the finite amount of coal in the earth and its essential role in the synthesis of all organic materials used by man, may persuade them to support options (1) and/or (2) instead. The risks involved in greatly expanded coal mining operations with its attendant human fatalities may also be a factor against option (3) even though it is least expensive.

In option (2), several negative factors must be considered. In the first place, enormous land areas are needed. For solar energy generation, there may not be sufficient quantities of arsenic, gallium, indium, and selenium to make all the solar cells needed to provide 1 TW (terawatt) of year-averaged power on 25,000,000 acres ($100,000 \text{ km}^2$) of sunny desert lands. The toxic nature of some of the solar cell ingredients, particularly if used in kiloton quantities, pose a health hazard (worse than uranium), and safeguards are needed to protect the public. In the case of wind power generation, the 2,500,000 large turbines (at 2 MW(e) each) placed on 30,000,000 acres ($120,000 \text{ km}^2$) of windy prairies to generate a year-averaged TW(e), will kill thousands of birds and spoil the landscape. Actually it is problematic whether such large numbers of solar panels and/or wind-turbines can be manufactured and installed in a 25-year time-frame by the year 2030 when oil shortages become severe. Regarding biofuels, if unaided by nuclear electricity, not enough land is available to grow enough biomass fuel for substitution of fossil fuels. Finally, the capital costs and operating costs of energy derived from solar and wind-power systems are four times higher than that for nuclear power. Despite these severe negative factors, some environmentalists may still prefer the 'Renewables-Only' option (2) over option (1) because of fear of 'things nuclear'. However for the survival of us all, common sense must prevail over the fear of a few.

While solar and wind-power sytems certainly have a niche in small-scale energy applications (e.g. power for small remote villages), these prime energy sources cannot economically provide the large amounts needed for our heavy industries and future synfuels manufacture. If nevertheless option (2) were to be adopted over option (1) in the USA, it would be very difficult for US manufacturers to sell goods on the world market in competition with countries that adopt nuclear option (1), because of their four-fold higher costs of electricity. Many jobs would be lost as a result, and poverty would increase.

Option (1), the Nuclear Millennium, is technically and financially feasible. It is less expensive than option (2) by a factor of 4 to 5. Nations that adopt option (1) will become future providers of mass-produced synfuels and will replace the current oil-producing OPEC countries that control the world's major energy supplies. The main problem foreseen at present is with the logistics of installing the prime-energy supply part of option (1), that is the difficulty of fielding a thousand new breeder reactors by 2030, when oil is expected to become scarce and expensive. A compromise may have to be made that allows prime energy to be provided by coal-burning plants for an interim period until uranium breeders can take over. If 2030 is the target date for switching to the new synfuels age, the USA must build 40 new breeder reactors each year (360 per year for the world) for the next 25 years! This is an extremely ambitious program and probably will have to be spread over a longer 50-year time span, at 20 new reactors per year (180 per year worldwide) using interim coal-power. That is, to bridge time delays in implementing option (1), coal-burning power plants will have to be kept in service longer, producing undesirable emissions of carbon dioxide. This will have to be tolerated a while for the sake of survival until nuclear plants can replace all coal plants. To succesfully implement option (1), a dedicated, focussed, and skillful commander will be needed like admiral Hyman Rickover who forged the formation of an operating nuclear navy in an incredible six years, or general Leslie Groves who under the WW-II Manhattan Project managed to have the design, testing, construction, and operation of gigantic isotope separation plants and first-ever nuclear fission machines completed in four short years.

Our firm conclusion is that option (1) is optimum and imperative for overcoming the pending energy crisis. While we still have sufficient oil, actions must be taken immediately to ward off the collapse of our economy by 2030. The risks involved in executing option (1) are infinitely smaller than the certain deaths of millions of people that would result if the 'Do-Nothing' option (4) is followed. Had the US government followed the original plan in the 1950's and 1960's to install three hundred nuclear power plants by the year 2000, the USA would be in much better shape today to confront global warming and the impending out-of-oil energy crisis. Instead, after the Three-Mile-Island and Chernobyl reactor accidents, it caved in to public fear-mongers and affected a temporary moratorium which kept the total US reactor fleet to the present one hundred nuclear power plants. This short-sightedness will be (and already has been) the cause of many energy shortages in the near-future. Unless quickly remedied while there still is time, it will assuredly be the unraveling of present-day US lifestyles and civilization [Ref. 53].

1.3.   ORGANIZATION OF THE BOOK

In Chapter 2 a review of common misconceptions about nuclear energy is given in the form of questionable assertions (fables) and factual answers (facts). It was felt important to present this material at the beginning of the book instead of at the end. The answers in Chapter 2 have been embellished to include deductions given in later chapters and to summarize the book's main message that an immediate acceleration in new construction of *uranium-breeding nuclear power plants (not weapons) is essential for our long-term survival*. Chapter 3 which reviews global energy demands and resources is probably the most important part of the book. It shows the necessity and means of developing alternative portable synfuels to replace petrol when oil and natural gas are depleted. Chapter 4 discusses current and possible future automobile engines and portable synfuels required to power them. Chapters 5 through 7 constitute a primer on nuclear electric power. Chapter 5 discusses various aspects of nuclear reactor operations, while those concerned about nuclear radiation effects should read Chapter 6. Chapter 7 discusses international measures taken to prevent diversion of fissionable fuels for use in nuclear weapons, a subject known as 'nuclear non-proliferation'. Chapter 8 presents concluding remarks and a list of action items.

CHAPTER 2

NUCLEAR FACTS AND FABLES

Many misconceptions have entered the nuclear folklore in recent decades. Major fables propagated by opponents of nuclear power are summarized here, and countered with facts. These facts are based on studies and data published by professional societies, representing some 250,000 diploma-ed engineers from around the world. Factual statements are backed up by data in later chapters.

Fable (1): "Nuclear reactors are like nuclear bombs".

Fact: This half-truth is frequently suggested by newspaper journalists who have little or no background in science and engineering. It is as erroneous and flawed as assuming that nitro-glycerin medicine used by heart-patients is as dangerous as nitroglycerin used in explosives, or that dihydrogen-oxide (water) is a dangerous chemical that drowns many people and should be banned. The uranium in a reactor is dispersed through a collection of fuel elements through which a coolant passes that absorbs heat and drives turbogenerators when the uranium undergoes fission. Nuclear reactors are designed so fissioning rates due to neutron multiplications are balanced and controlled by neutron-absorbing "control rods", yielding steady heat production. In today's reactors, if the core gets too hot, thermal expansion of moderator/coolant reduces neutron multiplication ("negative reactivity"), *and the reactor shuts itself down automatically*. This happens even if control rods are accidentally stuck and not instantly inserted in the core as would normally occur if a pre-set temperature is exceeded. In other words, the reactor *always* shuts down if it gets too hot.

The design of a nuclear fission *weapon* is entirely different. It comprises two halves or four quarters of highly enriched nearly critical fissionable uranium or plutonium which when slammed together (e.g. by springs), cause supercritical neutron multiplication and sudden production of an enormous amount of fission heat. This heat instantly evaporates all bomb material. If detonated in the atmosphere, it induces a shockwave that overturns and destroys any object in its path within a radius of a few kilometers. The physical arrangement that can cause a nuclear weapon to explode is *totally absent* in a power reactor. It is physically impossible for a reactor to explode like a bomb, just as much as it is impossible for a nitroglycerin-carrying heart-patient to be ignited and explode.

Fable (2): "We don't need more nuclear power; there is plenty of natural gas, oil, and coal."

Fact: In the 1990's, demand versus supply curves of natural gas and oil (petrol) crossed over, as predicted by Hubbert [Ref. 2]. Increases in demand now exceed discoveries of new oil deposits, and with these trends, oil and gas will be in serious short supply by 2030. Coal, if substituted for oil and uranium to provide all global energy needs might last 160 years. If more breeder reactors are put into service, uranium and thorium can supply the world with electricity and synfuels for at least 1,500 years in place of oil, gas, and coal. Like petrol, coal-burning power plants produce enormous amounts of air pollution and emit globe-warming carbon dioxide gas. Once oil is depleted, coal is more valuable as raw material for making organic chemicals, and should not be burnt. Non-air-polluting nuclear plants presently produce 21% of all US electric power. To avoid global warming, they should replace coal-burning power plants. The latest NEI (Nuclear Energy Institute) cost figures in ¢/kWh for electricity are: 1.71 (0.45) for nuclear, 1.85 (1.36) for coal, 4.06 (3.44) for natural gas, 4.41 (3.74) for oil, where parentheses give fueling costs [Ref. 37].

Fable (3): "Nuclear power is not needed. "Free" renewable solar, wind, hydro, and geothermal power will do. The utilities and government should invest more in them."

Fact: A 1200 MW(e) nuclear plant (e = electric) at 85% capacity factor produces 8,800 million kilowatt-hours of electricity per year, compared to about 50 million kilowatt-hours per year from a large SOLAR-2 station generating 15 MW(e) when the sun shines, with a 38% duty cycle. Thus it takes *one hundred seventy-six* SOLAR-2 stations occupying 25,000 acres ($100\,km^2$) of land and an investment of 10 billion dollars, to replace *one* nuclear plant occupying forty acres of land, costing 1.8 billion dollars. Solar energy is *not* free. Energy production has three cost components: fuel, maintenance, and capital write-off. It takes large maintenance crews and vehicles to keep solar panels free from dust, rain stains, and bird droppings, and to replace panels eroded or damaged by sand from dust storms. Also many square kilometers of collection surfaces are needed, so capital investment and write-off costs for solar stations dwarf the fuel costs for equivalent nuclear electricity. With ten-year solar-cell replacement cycles, one finds hazardous chemical wastes in manufacturing silicon, gallium-arsenide, or copper-indium-diselenide solar cells (requiring toxic silanes, arsenic, etc. as raw materials), far exceed uranium fuel wastes, when one compares 176 SOLAR-2 stations with 1 nuclear plant, each producing a year-averaged 1000 MW(e).

Regarding wind power, one comes to similar conclusions. Five-hundred 2 MW(e) windturbines costing 1.0 billion dollars, put on 30,000 acres (120 km^2), could yield 1000 MW(e) of electric power at full capacity. However the wind is not always blowing, and typical capacity factors for windfarms are 20%. To provide 1000 Mw(e) steadily for a whole year, electric storage batteries are needed (adding costs) and five times as many wind-turbines must be in place. In short, one needs 2,500 wind turbines of 2 MW(e) at a cost of $ 5 billion + $ 2 billion for interim storage, or a total of $ 7 billion to provide 1000 MW(e) year around. The typical capacity factor of a nuclear plant is 85%, so a 1200 MW(e) reactor costing $ 1.8 billion can provide an average of 1000 MW(e) during a year. Besides the high cost of maintaining 2,500 windmills, wind-farms have the problem of killing hundreds of birds and spoiling nature's scenery.

In summary, while the sun and wind may provide free fuel, it is not steady and highly diluted compared to enormously concentrated, reliable nuclear fission energy. To deliver large quantities of solar and wind-generated electricity, great expanses of collection equipment are required which vastly increase maintenance and capital costs relative to nuclear power plants. Solar and wind farms are very useful in providing electricity for small communities in remote locations (e.g. Alaska) or for low-power applications. However they could not economically replace nuclear or coal-fired power plants to feed an industrialized city with sufficient energy for manufacturing steel, bridges, buildings, or to produce massive quantities of portable synfuels for our transportation fleets of cars, trucks, ships, aircraft.

As a final example, let us compare in round numbers what it takes to generate a total of one million equivalent electrical megawatts presently consumed in the US, using either wind, solar, or nuclear power. With solar power, one finds that one must build 176,000 advanced SOLAR-2 plants (each producing a year-averaged 5.7 MW(e)), costing $ 10 trillion to deliver an average of one million MW(e) year around. This is more than the US gross domestic product (GDP) of $ 9 trillion. By the numbers shown above, providing one million MW(e) of windpower constantly during a year, requires 2,500,000 wind-turbines of 2 MW(e) peak power at a cost of $ 7 trillion. Compared with 1,000 nuclear plants of 1200 MW(e) each, that feed 1 million MW(e) to the entire USA with 85% capacity factor at a cost of $ 1.8 trillion, it is obvious what capital investors will decide when choosing between $ 1.8, $ 7, or $ 10 trillion. Note there are 438 nuclear power plants worldwide and 103 in the USA. The latter provide 21% of all electric grid power in the US.

Hydroelectric and geothermal power generation are maxed out in the USA. Most suitable rivers have already been dammed to feed hydroelectric turbogenerators; in fact environmentalists want to dismantle some hydroelectric dams. In Cobb, California, a geothermal power plant generating 55 MWe in the 1960's, experienced large drops in steam pressure and after six years was shut down. Recent geothermal projects are more promising but only useful in a few locations for a few decades.

Fable (4): "We only have 30 years of uranium ore to sustain a world fueled by fission power. Coal reserves would last at least 120 years, so we should concentrate on coal power."

Fact: The oft-quoted 30-year limit on uranium availability is based on "burning" the fissionable 235 isotope of uranium (U-235) only. Breeder reactors have been developed at slightly higher capital cost than U-235 burners, that consume U-238 as well as U-235 via in-core conversion of non-fissionable U-238 to fissionable plutonium-239, after U-238 absorbs a neutron (Section 5.1.2). Since uranium ore contains 140 times more U238 than U-235, consumption of U-238 ($\rightarrow$ Pu-239) gives the world uranium-based electricity for $140 \times 30 = 4,200$ years, or about 1,260 years with 30% utilization. Thorium after neutron absorption yields fissionable U-233, giving 300 more years of nuclear energy.

Coal is an alternative raw-material source for making industrial hydrocarbons such as plastics. Presently, oil provides the raw chemicals for manufacturing plastics, in addition to supplying petrol for the world's transportation fleets. Since oil reserves will be gone in forty years, it would be foolish to burn coal to deliver electricity, when non-polluting uranium fission power is available to generate all needed electricity. Besides, coal burning emits globe-warming carbon dioxide and other air-pollutants. It should not be burned.

Fable (5): "We should wait for development of nuclear fusion which produces no radioactivity."

Fact: Fusion reactors *do* generate radioactive isotopes in containment materials due to neutron activation. They burn deuterium and tritium making helium and neutrons. Removable neutron-absorbing inner linings have been proposed for fusion reactor chambers and these will become highly radioactive. Making fusion viable for nuclear power generation is a much more formidable task than generating electric power from fission. The minimum plant size to extract energy from a controlled fusion reaction (a miniature sun) is a hundred times that of a uranium fission reactor. It is estimated it will take at least another fifty years of research and development before the first fusion power plant might be built. We have waited fifty years already for a net-electric-energy-producing fusion pilot plant, and the no-oil period is approaching fast. Clearly we must proceed now with the expansion of proven uranium fission breeder technology.

Fable (6): "Hydrogen-consuming fuel-cell engines and electric energy storage batteries can replace petrol-burning automobile engines in the future; nuclear is not needed."

Fact: To be able to replace all present petrol-burning auto engines with fuel-cell engines, will depend on nuclear electricity or heat to produce the massive quantities of gaseous hydrogen (H_2) fuel needed for these new engines. Fuel-cell enthusiasts neglect to mention that H_2 gas is not a primary earth resource like oil, and must be manufactured. One needs electricity or heat from power plants to make lots of hydrogen. In effect this means that non-portable nuclear energy is transformed into portable hydrogen energy. Worldwide replacement of petrol-burning engines with fuel-cell engines is thus dependent on large-scale H_2 production obtainable only via nuclear- or coal-based power or heat.

There are at least five practical propulsion systems that could replace present automobile engines when oil is depleted. These are: (a) Combustion engines burning synthetically made fuels (synfuels) instead of petrol; (b) Hydrogen-consuming fuel-cell engines; (c) High-energy flywheels; (d) Electric battery packs, (e) Steam engines. Solar- and wind-driven cars are fun but cannot transport large numbers of people and goods. Future cars, trucks, ships, trains, and airplanes will most likely be propelled by synfuel-burning internal combustion engines (ICEs) or hydrogen-consuming fuel-cell engines (FCEs).

Large-scale hydrogen fuel production to replace all petrol presently used in transportation fleets, can be achieved by electrolysis or chemical reduction of water (H_2O) yielding hydrogen (H_2) and oxygen (O_2). The electricity or heat needed for this can be provided by nuclear power plants for more than fifteen hundred years or by coal-fired plants for over a hundred years. H_2 gas might be piped through gas pipelines to people's garages where it can be compressed in high-pressure cylinders to be carried on-board automobiles, or stored in a H_2-adsorbing "bladder", possibly the vehicle fuel-tank of the future. Alternatively, utility tap water might be electrolyzed to hydrogen and oxygen with available electric grid power in people's garages during the night. A porous fuel bladder can suck up hydrogen gas by adsorption on its inner surfaces and release it when slightly heated. ICEs or FCEs using oxygen from the air and H_2 fuel to move pistons or make electricity, exhaust only water. Thus water for making H_2 fuel (using electricity) is returned to water in a fuel-cell's exhaust. This is an eco-friendly cycle compared to globe-warming carbon-dioxide exhausts from petrol-burning ICEs.

If H_2 pipeline delivery or production in garages proves to be too expensive or unsafe, H_2 can be picked up at public fueling stations instead, either to refill hydrogen-adsorbing bladders or to replace liquid or compressed-gas cylinders filled with H_2. Proposals to bio-engineer H_2-producing organisms may be useful for low-quantity applications, but could never provide enough H_2 for all the transportation fleets in the world, as is generally the case for all biomass energy conversion concepts.

At present the main obstacles to the large-scale introduction of clean H_2-fuel-cell-powered cars is the H_2 storage problem and fuel-cell electrode fouling. Electric car engines have been developed, but replacement of the space presently occupied by an automobile fuel tank with the best H_2 adsorbing bladder or compressed-gas tank, results in a vehicle that can be driven for only one hour or 100 km (60 miles). Present techniques for H_2 bladder storage or compression need therefore a five-fold density increase to make H_2-fuel-cell-powered cars competitive with present-day petrol-fueled autos. Progressive fouling of fuel-cell electrodes may require their periodic replacement, like worn spark-plug replacements in today's combustion engines. This may be acceptable if the costs are reasonable.

Should the development of fuel-cell engines for automobile propulsion prove difficult, use of today's ICEs might be continued for a while after oil and gas reserves are gone, by fueling them with manufactured portable "synfuels" instead of petrol. With assistance of electric power, coal and water can be converted to syn-petrol (synthetic petrol) as is presently done in South-Africa's SASOL plant. Other synfuels producible with the assistance of electricity are hydrazine and ammonia synthesized from air and water, and ethanol obtained from sun-grown corn. To be efficient, the energy packed into a portable synfuel should not greatly exceed the amount of electric energy needed for its manufacture, although some energy conversion losses are justified. A nuclear plant does not fit in a car of course. To make abundant uranium-generated energy available for automotive uses, some losses are acceptable when this energy is converted and locked up in a portable synfuel or hydrogen. In making ethanol from sunshine and corn, electricity is used to manufacture fertilizers and farm machinery, and for husk removal, fermentation, and distillation operations. The combustion energy of ethanol (alcohol) is close to the electric energy required to make it. In a corn-alcohol-only economy this is unacceptable, but aided by nuclear electricity it is sustainable.

Use of methane (CH_4) or ethanol (C_2H_5OH) in a combustion engine still produces undesirable carbon dioxide (CO_2) emissions, while hydrazine (N_2H_4) or ammonia (NH_3) synfuels (e.g. for aviation) may generate unhealthy NO_x gases. Even if pure H_2 is used as synfuel in a combustion engine, high temperatures cause formation and exhausts of NO_x from reactions of oxygen (O_2) with nitrogen (N_2) in the air intake. Already developed catalytic NO_x converters or scrubbers might remedy the NO_x problem, but non-NO_x-producing fuel-cells operating at lower temperatures are preferred if practical.

Electric storage batteries and flywheels are other possible means of providing automotive power. However the most advanced flywheel systems and lightest battery packs developed to date are only able to provide enough energy to drive a small car for one hour. Flywheel or electric storage systems face the problem of diminishing returns: more energy storage to achieve a longer driving range means more battery or flywheel mass, which means more batteries and flywheels, etc. Unless a major breakthrough occurs that increases the kilowatt-hours/kilogram capacity of batteries and flywheels five-fold, it appears at present that fuel-cells and synfuels are the most promising for empowering the next generation of engines for mobile vehicles.

Fable (7): "Coal-fired power plants seem more bio-friendly than nuclear power plants."

Fact: Coal power requires coal transports using hundreds of railroad cars and release of tons of carbon-dioxide and natural radioactive elements into the atmosphere.

A 1000-MWe coal-fired power plant which consumes 4 million tons of coal per year, releases annually 900 pounds of coal-entrained uranium, in addition to 530 pounds of mercury, 120 million pounds of SO_x, 59 million pounds of NO_x, and 22 billion pounds (= 11 million tons) of CO_2 gas into the atmosphere.

Instead of dispersion through the atmosphere, nuclear-plant-produced radioactive waste is solid and contained. It is ultimately placed in some underground repository. A nuclear power plant is refueled once every 1.5 to 3 years with uranium encapsuled in solid fuel elements which are shipped in a few trucks. During refueling, the burnt-out "spent" fuel elements containing internal fission product wastes are removed and replaced with fresh fuel. After a brief cool-down period in a pond, the spent fuel elements are placed in a few collision-proof "caskets" for transport to a permanent nuclear waste site like the Yucca Mountain facility in Nevada under construction by the US Department of Energy. Here, after removing non-radioactive and short-lived radioactive species, long-life radio-isotopes are concentrated and stored in special nickel alloy containers, placed in underground caverns.

When uranium fissions, its energy is conducted through solid material to heat adjacent water (or gas) that runs the steam (or gas) turbines. In this heat transfer, radioactive fission products stay in the solid fuel elements, in contrast to coal burning, where species embedded in coal are sent into the atmosphere when coal is burning with oxygen in the air. Fissioned uranium products cannot undergo further fission or explode. They produce only low-level heat from radioactive decay. Even if a spent-fuel element were exposed to air during its transport in a collision-proof casket (e.g. if a terrorist fired a bullet into the casket), the solid form of radioactive products in fuel elements prevents their entry into the air. The heavy steel caskets are designed to tolerate external bomb blasts. It would take a casket-piercing missile with high explosives to vaporize a fuel element into radio-active aerosols.

Comparing safety in mining of coal versus uranium shows that many more accidents with loss of life occur in coal mines. Also daily railroad transportation of kilotons of coal are more accident-prone than monthly uranium transports of kilograms of uranium yellow-cake with a few trucks.

Fable (8): "Nuclear reactor operations are unsafe."

Fact: Two "maximum credible" nuclear power plant accidents involving core meltdowns occurred in the last fifty years, one at Three Mile Island (TMI) and the other at Chernobyl. They proved the soundness and safety of US and Western-Europe designed reactors, while they high-lighted the poor regard for safety and accident prevention under the former USSR regime. The Chernobyl power reactor had no heavy steel and concrete containment vessel as required in nearly all other countries, and was housed in a hangar. It also used graphite (very pure carbon) as moderator, which has a positive temperature coefficient of reactivity. In layman's terms this means that when the Chernobyl reactor core accidentally heated up beyond the control level, it promoted increased uranium fissioning that can cause a run-away power surge followed by a meltdown, unless halted by insertion of neutron-absorbing control rods. In contrast, in the USA and Western Europe, civilian reactors use water as moderator and coolant which has a negative coefficient of reactivity. When such a reactor gets too hot, the chain reaction terminates and the reactor shuts itself down.

The TMI accident happened because operators mistakenly forced it to overheat (thinking they were lowering the power level), causing the core to partially melt. However the safety features designed in the water-moderated TMI reactor fulfilled their function. The containment vessel held all radioactive core material in place. Except for minor escapes of tritium gas, no nuclear fall-out occurred. In the Chernobyl accident, maintenance technicians pulled out control rods in error, inducing runaway fissioning in the reactor core. The graphite moderator (a form of coal) got very hot and started burning with oxygen from the air as in a coal fire, because there was no containment vessel and unrestricted inflow of air. Firemen who had never been briefed about nuclear reactors tried to put out the fire but unknowingly exposed themselves to lethal levels of radiation. The 3 maintenance technicians instrumental in starting the Chernobyl accident were instantly killed by flying debris, while 28 firemen and rescue-workers died from radiation overdoses within months [Ref. 34]. Heart attacks killed 3 more, while 11 succumbed from medical complications years later. Another 30 rescue-workers exposed at the Chernobyl site suffered permanent disabilities.

The fear of nuclear power plant accidents seems irrational when compared to air and car accidents. Air and car crashes kill thousands of people each year, yet few people want to abolish cars and airplanes. In the past fifty years, less than a hundred people worldwide died in nuclear accidents, even though nuclear power provides 21% of all electricity in the US and 85% in France. To produce clean non-air-polluting electricity in the US, it is imperative that more nuclear power plants be built to replace coal- and gas-fired units. The latter will be inoperable when gas reserves are depleted.

Fable (9): "We don't know what to do with "dangerous" radioactive waste from nuclear reactors."

Fact: We *do* know what to do. The *annual* fission product waste from *all* 103 nuclear power plants in the USA, which produce nearly *one trillion* (10^{12}) kilowatt-hours of electricity per year, can be extracted, concentrated, and compacted as ceramic marbles in a few hundred drums, to be stored underground in the national nuclear waste repository at Yucca Mountain, built in the Nevada desert. Because of anti-nuclear politicking, completion of the Yucca facility which was supposed to have been ready by 2000, has suffered delays. Even though collision-proof caskets will be used which have been crash-tested extensively, transportation of nuclear wastes over US highways and railroads is still opposed by anti-nuclear activists.

This has forced nuclear plant operators to temporarily store used fuel elements in water-shielding swimming pools until Yucca is operational. If properly prepared, temporary swimming-pool storage of spent fuel elements is safe. But it is still better to store the waste in one place rather than on a hundred different sites. Aside from civilian nuclear power plants, the US Nuclear Navy has similar spent-fuel loads to dispose of each year. There appears to be a misconception among nuclear-power opponents that "dangerous" radioactive waste can somehow explode like uranium through nuclear fission. *Radioactive waste cannot explode and is absolutely non-fissionable. It only suffers from slow nuclear decay*, which entails emissions of betas (= fast electrons) and gammas (= high-energy photons similar to x-rays).

Fable (10): "The longer the lifetime of a radioactive element, the more dangerous it is for man."

Fact: Just the opposite is true. The intensity of radiation from a gram of radioactive material is lower the longer its decay lifetime. Conversely it is higher, the shorter its life is. We are surrounded by natural long-lifetime radioactive materials on our planet. In fact, each human is internally radioactive because of the potassium (K) present in every human cell. Natural potassium has 0.12% radioactive potassium-40 (K-40) isotope in it. K-40 emits beta and gamma radiation and decays with a half-life of a billion years compared to a four billion year half-life for uranium-238 decay. A recent uproar in Europe over depleted uranium-238 used in military projectiles, shows the technical ignorance of "green" politicians who are easily brain-washed by anti-nuclear propaganda. Mildly radioactive "yellow-cake", a uranium oxide produced after uranium mining and pre-processing, is not a "nuclear explosive" as some mistakenly believe. It is as harmless as a potassium-carrying mineral or thorium-oxide mantle in a Coleman lantern. Besides unremitting exposure to internal K-40 radiation, man is bathed in natural radiation coming from cosmic sources and from the earth. He has evolved just fine with all this radiation. Recent studies show mild radiations may even be beneficial [Ref. 35].

Fable (11): "Thousands of people can die after a nuclear plant meltdown."

Fact: There have been two major nuclear reactor meltdown accidents since the beginning of the nuclear power era, one in the USA and one in Russia. The actual fatalities are 0 deaths in the USA from the Three Mile Island (TMI) accident, and 45 in the former USSR at Chernobyl in the Ukraine [see Section 6.6 and Reference 34]. I personally visited Chernobyl and the regional hospital near Pripyat, and talked to local residents and operators of the three Chernobyl reactors (only one had a meltdown). Shortly after the Chernobyl accident, scare-mongers predicted thousands would die later from fall-out radiation. This is total nonsense. Actual nuclear fall-out victims in the Chernobyl region were children who drank contaminated milk from cows that had eaten contaminated grass (avoidable if authorities had warned farmers). These children accumulated radioactive iodine in their thyroids. By administration of iodine-displacement therapy and waiting till the radioactivity subsided (Iodine-131 has an 8-day half-life), the affliction disappeared for most of them after a few months. Of an estimated 3000 people exposed to fall-out, 9 people were recently reported to have died, allegedly from exposure to Chernobyl's nuclear fall-out.

Claims of thousands of future cancers due to Chernobyl fall-out made by antinuclear groups are based on distorted probability calculations not acceptable to statisticians. Under-reported pre-Chernobyl cancers, and cancer cases due to modern chemicals which cannot be distinguished from nuclear-fallout-generated cancers, produce flawed statistics. As all mortals do, most of the 140,000 evacuated inhabitants in the direct fall-out path of Chernobyl's radioactive plume will die between ages 60 and 100. Based on world-wide cancer-death statistics, at least 14,000 ($\sim 10\%$) of them are expected to die from cancer *due to non-nuclear causes*. Antinuclear propaganda claims *all* these deaths will be due to Chernobyl, a totally untenable charge. It is akin to claiming that coffee kills 20% of all people, based on the fact that 20% of all people drink coffee and all will ultimately die.

Fable (12): "Exposure to "radiation" causes long-term after-effects in one's body."

Fact: The word "radiation" is repeatedly misused by lay people and substituted for radioactive particles (see below). In physics, gamma radiation from radioactive processes falls in the same class as visible light radiation, infrared heat radiation, and radio waves. All are made up of massless electromagnetic waves or evanescent photons which can be absorbed or reflected once, but do not "stick" as some people mistakenly believe. Like heat which emits infrared photons, a little bit of radiation is harmless and even beneficial (e.g. a heating pad), but too much can kill you (heat in an industrial furnace incinerates you). Nuclear reactor cores emanate alpha and beta particles, neutrons, and gammas. Emanations with mass such as the beta particles (which are fast electrons) and alphas (Helium ions) are stopped by less than a millimeter of metal or concrete, while neutrons are absorbed or reflected back into the reactor core. Only gamma radiation emitted by decaying fission products requires thicker stopping materials. Massless gamma radiation is like massless solar ultraviolet light, except the frequency and thus photon energy is higher. Reactors have enough shielding around them to absorb most massless gammas, allowing only an insignificant harmless number to get through.

A person's exposure to a beam of gamma photons emitted by a radioactive compound can cause breakage of a few biochemical bonds in body tissue. However the body does not differentiate between broken biomolecular bonds from a scratch, a knife-cut, cosmic radiation, or from gamma photons. A scratch may be more detrimental than gamma damage since broken bonds are closer together in a scratch, while molecular breakages due to gammas are spread out. People don't know what it means when told they have been exposed to "100 millirems of radiation". This number can be put in perspective, knowing body damage from a 2 cm long, 0.1 cm deep scratch on one's skin causes biomolecular bond breakages equivalent to 100 millirems. The human body repairs broken bonds rapidly and has done so during a million years of evolution in a radiation-rich environment.

A more important nuclear safety concern is inhalation or ingestion of radioactive particles or dust present in the "fall-out" plume of atomic bombs or in the debris cloud from the meltdown of a reactor without a containment vessel such as Chernobyl. Some body organs (e.g. thyroid gland) and bones have an affinity for certain uranium fission products, mainly radioactive iodine, cesium, and strontium. The body can extract these elements from inhaled or ingested radioactive dust, and concentrate them unless they are eliminated ("anti-radiation" pills are available today that can force the body to expel such undesirable elements). When lodged in the body they can constantly emit betas and gammas in surrounding tissue and irritate or destroy it (this is exploited in nuclear medicine to kill cancer cells, but here cancerous tissue is pre-selected).

In the unlikely event one is in the path of the debris cloud from an atomic bomb or Chernobyl-like explosion, the best protection against fall-out is to enter a shelter with closed windows. If outside, one should filter the air one breathes using a wet handkerchief or gas mask, and wash off all dust by taking a swim or shower after the cloud has passed. If available, one should take anti-radiation pills. Of course the best protection is to run or drive away from such a usually slow-moving cloud.

To avoid possible radioactive fallout entirely in nuclear melt-downs, today *all* power reactors in the world must have a steel and concrete containment vessel surrounding them. This vessel must keep all nuclear reactor debris contained under the worst imaginable ("maximum credible") accident such as a core meltdown, an M-8 earthquake, airplane crash, (non-nuclear) bomb attack, sabotage, etc. The (almost incredible) Three-Mile-Island (TMI) accident proved its effectiveness in limiting damage.

CHAPTER 3

ENERGY CONSUMPTION AND ENERGY SOURCES ON PLANET EARTH

3.1. DEFINITION OF ENERGY AND UNITS FOR ENERGY AND POWER

As taught in every highschool, it takes kinetic energy for anything to move. This movement energy can be obtained by conversion of stored-up (potential) energy that is released. Without energy and energy conversions, the whole Universe would be dead and we would not exist. Since we will be discussing energy usage, energy exchanges, and energy supplies, it is necessary we define a unit of energy. For example how many units of energy are in a liter or gallon of petrol.

The word 'energy' comes from the Greek meaning 'inherent work'. Although others before him had hinted at the conservation of mechanical work and heat, it was Sir James Prescott Joule (1818–1889) who first carefully measured and proved the inter-convertability of heat and mechanical work, firmly establishing the abstract concept of energy and conservation of energy. He showed that a certain amount of mechanical motion energy could be produced by a certain amount of heat energy. Also that a big car needs more energy to be moved than a small car, in proportion to its weight.

The laws of energy conservation and energy exchange are the cornerstones of physics. One can define energy on the microscopic as well as macroscopic scale. Microscopic atoms, molecules, electrons, protons, neutrons, nuclei, photons, etc. are all endowed with energy, in addition to mass, charge, etc. Likewise cars and trucks moving over a highway possess kinetic energy, acquired by converting petrol-fueled heat of combustion in their engines into mechanical motion of their wheels and thence onto their vehicle. People who drive cars can visualize energy best by equating it with liters or gallons of petrol. They know their car needs 60 liters or 16 gallons of petrol to fill their tank that allows them to drive 600 km or 373 miles. In technical parlance, the chemical energy contained in 60 liters (16 gallons) of petrol when liberated as heat of combustion, is converted by the engine to mechanical energy of wheel rotation, taking the car a distance of 600 km (373 miles). Thus one can equate 1 liter of petrol energy with 10 kilometers of mechanical work, or 1 gallon with 23 miles.

Energy can be in the form of kinetic energy (e.g. a falling stone) or potential energy (e.g. a stone on the edge of a cliff ready to fall), one being convertible into the other. Chemical energy stored in molecules like petrol, and nuclear energy present in atomic nuclei, are both forms of potential energy that can be converted into kinetic energy under certain conditions. Heat is the total kinetic energy from swarms of chaotically moving or vibrating molecules or atoms. Heated molecules in a gas can be directed to push a piston, thereby converting heat into mechanical energy of motion. When hydro-carbon ($C_m H_n$) molecules in petrol react with heated atmospheric oxygen (O_2) in a combustion engine, C, H, and O atoms are rearranged into new molecular compounds (CO_2 and H_2O) with liberation of kinetic energy in the form of heated gases that move pistons. Similarly a neutron flying into the nucleus of a uranium atom, can cause a re-arrangement of protons and neutrons in the nucleus (Chapters 5 and 6). This results in the splitting (fissioning) of a uranium nucleus into two halves and liberation of kinetic energy imparted to the two recoiling fission fragments which generate heat in the solid that embeds them. The amount of energy liberated in the fission of a nucleus is generally ten million times larger than that liberated in a chemical reaction. This is the reason why a nuclear plant can produce so much more power from a kilogram of nuclear fuel (uranium), than a coal- or oil-fired power plant can generate from a kilogram of petro-chemical fuel.

The physicist's unit of energy is aptly called the Joule (abbreviated J). Power is defined as the energy delivered per unit time or the energy *rate*. In physics, the standard unit of power is the Watt (W) or Joule per second (J/s). That is, $1\,W = 1\,J/s$. Comparing energy with a cup of water, then power is like cups of water pouring out of a faucet per unit time. The antiquated 'horsepower' (HP) unit, based on the strength of horses, is still used to rate car engines. It equals 746 Watt, that is $1\,HP = 746\,W = 746$ Joules per second. Historic definitions of various other units for energy and power can be found in physics textbooks. For example the calorie energy unit which is still used, is based on heating 1 gram of water by 1 degree centigrade (Celsius), and equals 4.2 Joules ($1\,cal = 4.2\,J$). For multiples of a basic unit, one uses k for kilo (thousand or 10^3), M for mega (million or 10^6), G for giga (billion or 10^9), and T for tera (trillion or 10^{12}). Thus 1 kW equals one thousand Watts of power, 1 MW is one million Watts of power, etc. A peculiar energy unit is the kWh or kilowatt-hour which is disguised as if it is a power unit. It actually is an energy unit and represents energy delivered at a rate of 1000 Watts = 1000 Joules/second for a period of 1 hour = 3600 seconds. Thus $1\,kWh = 1000(J/second) \times 3600$ (seconds/hour) $= 3.6$ million $J = 3.6\,MJ$.

In dealing with large quantities of energy, we shall use three units, the Giga-Joule $= 1\,GJ = 1$ Billion Joule; the MegaWatt-hour $= 1\,MWh = 1$ Million Watts

ENERGY CONSUMPTION AND ENERGY SOURCES ON PLANET EARTH

for One Hour; and the MBTU = 1 Million BTU (British Thermal Unit). They are related as shown in Brief 1[2]. Conversion factors are needed so that one can compare reported data from energy sources which use different units. In comparing amounts of energy generated from a kilogram of oil, coal, or uranium, it is also important to specify whether the energy is in the form of heat, electricity, or mechanical motion. We follow the convention of placing (e) or (m) in parentheses after units of energy for the latter two; otherwise it is assumed to be heat. Thus 1 GJ(e) designates electric energy, while 1 GJ is a quantity of heat. The distinction is important because electrical and mechanical energy are of a higher grade.

Most of man's energy usage involves mechanical motion or electricity, obtained by conversion of heat energy via a steam or gas turbine, or by direct electrochemical energy conversion to electricity. Physics shows that usually only 30% to 40% of heat (= chaotic molecular motion) can be converted by a turbine into macroscopic mechanical motion or electricity[3]. The balance is dumped as low-temperature heat into the air or coolant (lake or ocean water). Direct conversion of chemical energy into electricity (fuel cells) and mechanical motion is more efficient, and can take place with (practical) efficiencies of 45% to 85%. In evaluating world energy resources, fuels are measured by mass (volume and density) whose energy content is given as latent heats of combustion, while hydro, wind, or solar sources specify electric outputs. In comparing these energy forms, we shall assume a coarse conversion factor of 33% for conversion of heat into electricity. For electrochemical conversions we shall assume efficiencies of 55%, while for interconversions of mechanical and electric energy we assume ~100% (Brief 1).

It is helpful to recall here that a tankful of petrol in a medium-sized automobile contains about 60 liters (16 US gallons). This contains 2.2 GJ of chemical combustion energy which can be converted to mechanical motion to propel the car a distance of approximately 600 km (373 miles). Conversely 1 GJ of heat energy is stored in 27 liters (7.3 gallons) of petrol which moves a car 273 km (170 mi).

Brief 1 lists the approximate equivalences for automobile propulsion and travel using petrol, ammonia, and hydrogen fuels. Heat of combustion in an ICE is assumed to be convertable to mechanical energy by a factor of 0.33, while the efficiencies of FCEs are assumed to be 0.55. Ammonia and/or hydrogen may become the main fuel of the future for automotive fuel-cells (Chapter 4).

---

[2] We shall use the word 'brief' for any table, chart, sketch, or figure that summarizes or illustrates a concept or relationship.
[3] The 'second law of thermodynamics' states that the maximum mechanical energy extractable from heat is given by the Carnot fraction $(T_1 - T_2)/T_1$, where T_1 and T_2 are inlet and outlet turbine/engine temperatures.

BRIEF 1: ENERGY AND POWER UNITS AND CONVERSION FACTORS

ENERGY

$1\text{ GJ} = 10^9 \text{ J} = 0.278 \text{ MWh} = 278 \text{ kWh} = 9.48 \times 10^5 \text{BTU} = 0.948 \text{ MBTU}$

$1\text{ MWh} = 1000 \text{ kWh} = 3.6 \text{ GJ} = 3.413 \times 10^6 \text{ BTU} = 3.413 \text{ MBTU}$

$1\text{ MBTU} = 10^6 \text{ BTU} = 1.055 \text{ GJ} = 0.293 \text{ MWh} = 293 \text{ kWh}$

POWER:

$1\text{ GJ/y} = 31.71 \text{ J/s} = 31.71 \text{ W} = 0.03171 \text{ kW}$

$1\text{ W} = 1 \text{ J/s} = 3.6 \text{ kJ/h} = 31.54 \text{ MJ/y} (1\text{ y} = 3.154 \times 10^7 s)$

$1\text{ kW} = 1 \text{ kJ/s} = 3.6 \text{ MJ/h} = 31.54 \text{ GJ/y}$

HEAT → ELECTRICITY OR MECHANICAL ENERGY:

$3\text{ GJ} \rightarrow 1 \text{ GJ(e, m)}; \; 3\text{ MWh} \rightarrow 1 \text{ MWh(e, m)}; \; 3\text{ MBTU} \rightarrow 1 \text{ MBTU(e, m)}$

EQUIVALENCES FOR INTERNAL COMBUSTION ENGINES (ICE):

One 'tankful' petrol ≈ 45 kg petrol ≈ 60 liters(16 gal) of petrol

≈ 2.2 GJ $\rightarrow 0.72$ GJ(m) $\rightarrow \sim 600$ km(373 mi) of car travel

One 'tankful' ammonia ≈ 47 kgNH$_3$ ≈ 40 liters(11 gal) of liquid ammonia @ 20 atm

≈ 2.2 GJ $\rightarrow 0.72$ GJ(m) $\rightarrow \sim 600$ km(373 mi) of car travel

EQUIVALENCES FOR FUEL CELL ENGINES (FCE):

One 'tankful' hydrogen ≈ 10 kgH$_2$ ≈ 600 liters(160 gal) hydrogen gas @ 245 atm

≈ 1.3 GJ $\rightarrow 0.72$ GJ(e, m) $\rightarrow \sim 600$ km(373 mi) of car travel

One 'tankful' ammonia ≈ 28 kgNH$_3$ ≈ 24 liters(6.5 gal) of liquid ammonia @ 20 atm

≈ 1.3 GJ $\rightarrow 0.72$ GJ(e, m) $\rightarrow \sim 600$ km(373 mi) of car travel

---

3.2. AMOUNTS AND FORMS OF ENERGY CONSUMED BY MAN

According to statistics supplied by the US Census Bureau and Department of Energy (DOE), there were 281,422,000 people living in the USA in 2000, who consumed a total of 1.18×10^{11} GJ/y of heat-equivalent energy from the primary energy sources listed in Brief 2A. The U.S. consumption rate was thus 419 GJ/y or 4.4 kW(e) per person. This compares with a total world consumption of 4.01×10^{11} GJ/y of heat energy by 6,157,401,000 people, or 67 GJ/y (0.71 kW(e)) per person. These per-capita consumption figures might indicate that a US resident is consuming 6 times the world average. However a lot of hardware in the world (cars, planes, ships, bridges, tractors, etc.) used in non-US countries were fabricated in the USA, so some of the energy for their manufacture must be allocated to non-US residents. This increases the 67 GJ/y figure and decreases the US figure of 419 GJ/y. These considerations apply primarily to Asia, Africa, and South-America who buy such hardware in exchange for labor-intensive (non-petrol-consuming) goods, oil, and raw materials. Europe and Japan, like the USA, also make energy-consuming hardware products traded throughout the world. Without considering detailed balances of world trade and energy exchanges, coarse estimates change the above figures to about 73 GJ/y or 0.77 kW(e) per person for the world and about 300 GJ/y or 3.2 kW(e) per US citizen, still four times the world average.

In the next 20 years, one can expect non-USA energy consumption to increase, particularly in China. The world consumption rate is estimated to increase to $123 \text{ GJ}/y = 3.9 \text{ kW} = 1.3 \text{ kW(e)}$ per person when averaged over the years between 2005 and 2025. With a world population of 6.1 billion in 2000 leveling to 7.8 billion predicted for 2025, these figures forecast a world energy consumption rate of about 0.86 trillion (10^{12}) GJ per year averaged over the next 20 years. With total primary energy reserves as listed in Brief 2B, one then calculates depletion times of 16.4 years for oil, 18.4 years for natural gas, 153 years for coal, and 1100 years for uranium, *assuming* all needed heat-equivalent energy is supplied *only* by oil, or *only* by natgas,[4] or *only* by coal, or *only* by uranium. Exploitation of 1.5 trillion barrels of oil from shale and tar-sands, and 10 quadrillion cubic feet of natgas

BRIEF 2A: ANNUAL ENERGY RESOURCE CONSUMPTION IN THE USA (2000)			
Energy Resource	Annual Quantity Consumed	Equivalent Heat Consumption	Percentage
Oil	7.08×10^9 barrels/y	4.0×10^{10} GJ/y	33.90%
Natgas (Natural Gas)	2.38×10^{13} cu.ft/y	2.5×10^{10} GJ/y	21.18%
Coal	1.70×10^9 tons/y	3.74×10^{10} GJ/y	31.68%
Uranium[1]	20,000 tons/y[1]	0.85×10^{10} GJ/y[1]	7.20%
Hydroelectric[2]		0.33×10^{10} GJ/y	2.80%
Geothermal[2]		0.034×10^{10} GJ/y	0.28%
Wood/Bio, Wind, Solar[3]		0.35×10^{10} GJ/y	2.96%
TOTAL:		11.80×10^{10} GJ/y	100%

NOTES: [1] With present U-235 'burners', only 0.5% of the intrinsic uranium energy is utilized. With U-238 breeder reactors, only 100 tons/y would be needed to provide 0.85×10^{10} GJ/y; [2] Hydro and Geothermal are close to the maximum available in the USA; [3] Wind and Solar may expand three-fold in the next twenty years but most likely can never provide more than ten percent of total energy needs.

---

[4] We shall abbreviate 'natgas' for natural gas (mix of methane, ethane, propane, butane) from here on.

BRIEF 2B: PRESENT (2004) WORLD RESERVES OF PRIME ENERGY RESOURCES					
Resource	Quantity	Heat Content	Conversion Factor	Depletion Time @ 123 GJ/y per man (Popul'n = 7×10^9)
Oil (including tarsands)	2.5×10^{12} barrels	1.41×10^{13} GJ	5.65 GJ/barrel	16.4 years
Natgas (including sea-beds)	1.5×10^{16} cu.ft	1.58×10^{13} GJ	1.05 GJ per 1000 cu.ft	18.4 years
Coal	6×10^{12} tons	1.32×10^{14} GJ	22 GJ/ton	153 years
Uranium	1.1×10^7 tons	8.60×10^{14} GJ (^{235}U + ^{239}Pu)	8.6×10^7 GJ/ton	1100 years
Thorium	3×10^6 tons	2.87×10^{14} GJ (^{233}U)	8.6×10^7 GJ/ton	333 years
TOTAL:		9.63×10^{14} GJ		

NOTES: Depletion times assume all mankind's energy needs are provided by one resource only. 1 barrel = 42 gallons = 159 liters; 1 gallon = 3.785 liter; 1 cuft = 28,316 cm^3 = 28.316 liters; 1 ft = 30.48 cm; 1 mile = 1.609 km; 1 lbs = 0.454 kg; 1 ton = 2000 lbs = 907.19 kg; 1 tonne = 1000 kg = 2204.62 lbs; 1 year = 365 days = 8,760 hours = 525,600 min = 3.154×10^7 sec.

from sea-beds, requiring less than 20% of contained fuel energy for recovery, are included. If tar-sand oil and sea-bed natgas are excluded, only 1 trillion barrels of oil and 5 quadrillion cubic feet of natgas are left, and depletion periods change to 7 years for oil and 6.5 years for natgas. Of course it is unrealistic to assume that *only* oil, *only* natgas, *only* coal, or *only* uranium will be used to support *all* of man's energy needs. Nevertheless these depletion periods are useful to indicate the relative mortality of these resources, and to show uranium's superior long life.

In the real world, the transportation sector which comprises our vast fleets of land, air, and sea vehicles, consume most of the available oil. In the USA, about 35% of all energy is consumed by transport vehicles, but world-wide this percentage is closer to 40% (the balance of 60% is mostly electricity), since less energy for heavy industry is used. At 3.44×10^{11} GJ/y, the actual availabilty of petrol beyond the year 2005 would then be 24 years without tar-sands oil, and 41 years with tar-sands oil included.

While locomotion of most transportation vehicles is obtained via petrol-burning combustion engines, electricity is generated by means of steam or gas turbines that utilize heat obtained from coal, uranium, or natgas. Additional small quantities of electric energy are provided by geothermal sources, hydroturbines, windturbines, and solar cells. To prolong the epoch of the well-developed combustion engine, natgas can be compressed (at about 120 atm) and used in place of petrol. In The Netherlands where the price of petrol is four times higher than in the USA, many automobilists used cylinders of compressed natgas instead of petrol in the 1970's when natgas was less expensive. That is, compressed natgas can be substituted for petrol to propel automobiles when oil becomes scarce and expensive. For this reason, it would be prudent to preserve natgas for future use as a portable fuel and not waste it now in elecric power generation, which can easily be run on coal and uranium alone. If we assume that all of the presently available 5 quadrillion cubic feet of natgas (*not* including speculative retrieval of methane-hydrate from sea-beds) will be available for fueling our transport vehicles, in addition to the 2.5 trillion barrels of oil (including tar-sands oil), the period for continued use of the well-developed combustion engine might be extended. Instead of 41 years, it then would take 56 years to petro-fuel exhaustion. This is with the proviso that oil and natgas are used *only* for vehicle fuels, and electricity is produced *only* by coal, uranium, and renewables. Under this scenario, more time is available to develop new propulsion systems and synfuels for aircraft, ships, and land transport. Many believe the above time estimates are too optimistic and that the most-probable out-of-oil and -natgas time-point will be reached in 40 years.

If the burning of coal in electric power plants is halted to reduce global warming and to use/conserve it as a raw material for making plastics,[5] one finds that present nuclear power generation must be expanded sixfold to replace all coal power plants presently used for making electricity. For the USA, this means that five-hundred uranium-burning plants must be added to the existing one hundred operating nuclear power plants. However a further expansion of uranium power plants from five hundred to one thousand units of 1200 MW(e) each must be in place to accomodate the manufacture of portable fuels with nuclear electricity or heat for new vehicle propulsion systems that can no longer depend on petrol or natgas. Severe oil shortages will develop well before the nominal 40-year total depletion period is reached, as production from different oil fields are reduced or stopped (prices increased!) when they approach exhaustion. It is estimated this will happen in about 25 years. Though it is impossible to set a precise date, we believe *we will see few petrol-driven cars after 2030*. As indicated in Brief 2, electricity from 'renewables' (hydro, wood/biomass, geothermal, wind, solar) helps. But for them to produce enough portable synfuels for all the world's vast transportation fleets is virtually impossible (Subchapter 3.3).

In the manufacture of portable synfuels there are losses in converting prime heat or electricity into chemical energy of a synfuel. As shown in Chapter 4, synthesizing hydrogen, ammonia, or hydrazine from air and water actually can be done only with an efficiency between 10% and 60%. That is, 40% to 90% of prime energy is lost in converting it into portable synfuel energy. As long as there is an abundant source of prime uranium (or coal) energy, this poses no problem. Even if it would take 3 GJ of prime energy to make 1 GJ of synfuel energy, there is no bottleneck. To a traveling automobilist, portable synfuel energy is more valuable than non-portable nuclear reactor heat. He does not mind if a substantial amount of the original nuclear

---

[5] The word 'plastics' is used here to include carbon nanotubes, fibers, and most materials and products presently derived from petrochemicals.

heat used for chemical synthesis is lost, just like he is quite willing to waste 66% of the heat produced in his internal combustion engine as long as the balance of 33% is converted to move his vehicle. On the other hand if the only source of prime energy is biomass-generated alcohol, and it takes more alcohol fuel to grow alcohol fuel (due to running of farm equipment, distillation process, etc), the situation would be unsustainable in a no-oil, no-coal, no-uranium future. Only with uranium-generated electricity or heat to provide the energy needed for cultivation of plants and to extract their alcohol, can bio-alcohol be a practicl synfuel, converting non-portable nuclear energy into a portable fuel.

Electricity is one of the greatest gifts to mankind allowing him to communicate by telephone, radio, television, and to have all the comforts of the modern home such as electric lighting, air-conditioning, refrigerators, heating, cooking, etc. Without electricity we would have candles and torches for lighting, cold water for bathing, spoiling food, and sweltering dwellings in summer. Instant communications around the world would also be impossible. It is very fortunate that nature has provided us with very light negative electrons that can pass swiftly through metal conductors such as copper. If electrons would have been heavy (with the same mass as positive protons), there never could have been readily available inexpensive electricity as we know it. Because of the properties of electrons, electric power can be delivered rapidly and distributed widely with little propagation loss, to the great benefit of man. The same comment applies to uranium fission which allows the entire world to have at least a thousand years of electric energy. In the early days of electric power, protesters tried to block its distribution, claiming that thousands of people would die if high-voltage AC power lines would be stretched out over the land. Similar preposterous assumptions are made by today's anti-nuclear lobbyists who try to impede expansion of nuclear power. Senator Robert Kennedy once observed that in almost every national issue 'One-fifth of the people are always against,' and that the contrarians are quite bull-headed. Philosophically one can argue whether electricity and nuclear power are a blessing or curse to man. It is up to man to use these gifts of nature for good or evil. One hopes destructive uses can be rooted out and all of mankind will band together to enjoy the benefits of ample uranium-generated electricity for millennia.

3.3. LIMITATIONS OF 'RENEWABLE ENERGY'

Recently there have been assertions that nuclear energy is not needed and that all oil-derived petrols can be replaced by 'renewable' bio-mass fuels such as corn-derived alcohol, bio-diesel, bio-hydrogen, etc. We shall examine this proposition now in some detail. As discussed, the estimated world-averaged energy consumption rate between 2005 and 2025 will be about 1.3 kW of equivalent electric energy per person, or 9.1 billion kW for 7 billion people[6]. A typical 100-Watt lightbulb, when

---

[6] In this Subchapter, we shall omit (e) in kW(e) for brevity.

turned on, continuously burns 100 Joules of electric energy per second. If in the next two decades each person consumes 1.3 kW of energy on average, he/she would continuously burn the equivalent of 13 electric lightbulbs. This energy consumption rate includes each person's share of oil and electricity used in making foods and goods and their transport to market, for fuel to drive to work and back, for making steel and aluminum used in cars, bridges, buildings, ships, airplanes, appliances, and for home lighting, cooking, heating, cooling, etc. The world-average figure of 1.3 kW per person (presently 0.77 kW) assumes that standards of living in China, India, Indonesia, and other countries will improve in the next decades. Also it is assumed that reasonable conservation will be practiced and excessive energy is not wasted in too many global scorched-earth wars.

Solar radiation at the earth surface amounts to about 1.35 kW per square meter or 5,463 kW per acre when sun-shine peaks. The overall efficiency of plants or trees to convert earth-incident solar energy into burnable carbonaceous chemical energy (alcohols, wood, etc) is about 0.03% for the best bio-fuel producers [Refs. 48, 49]. This includes average diurnal, seasonal, and weather effects, as well as foliage intercept fractions, crop turn-over times, and plant photosynthesis efficiencies (plants use a lot of captured solar energy for pumping water). With 5,463 kW per acre of peak solar irradiation, plants might thus produce 1.6 kW per acre of potential bio-fuel. Assuming 10% for access roads to cultivate and harvest, one acre could yield 1.5 kW of bio-fuel energy, if crops are continuously replanted after harvesting. We are talking here about a continuous balance between solar energy delivery, conversion, and harvesting of bio-energy. One can accumulate bio-energy for years by growing timber for example, and consume it all in a few hours at a thousand kW (a million Joules per second) per acre in a forest fire. But it takes a long time for re-growth before a repeat of such a high energy release rate is possible. Because of energy conservation (what goes out must come in), one can only extract 1.6 kW per acre of year-averaged bio-fuel energy. The fossil fuels we burn up today in one year, were deposited by plants over millions of years in the past.

To replace all the world's petro-fuel and electric energy with synfuels and electricity derived *only* from bio-mass, will take about (9.1 billion kW)/(1.5 kW/acre) = 6 billion acres of arable land. To grow, harvest, and process bio-crops, one needs tractors, fertilizers, processing operations, etc., which all consume energy. Proponents of biofuel production claim this takes less than 90% of the alcohol energy produced by corn crops for example. Assuming that improvements can lower this number to 80%, bio-fuel production would need five times more land, or 30 billion acres to generate a net of 9.1 billion kW of marketable portable bio-fuel. Since the world's total arable land is about 8 billion acres, this is four times more than what is available. The total surface area of the world's land mass is about 37 billion acres (the USA has 2.24 billion acres), much of which is permanently frozen. Clearly it is impossible to accomodate the world's energy needs if one depends *only* on bio-mass. However with the aid of nuclear electricity which can run alcohol distillation plants, fertilizer production, and the manufacture of farm equipment, four times less land is needed and "only" one-third of the USA's lands

(0.6 billion acres) is needed. In this case, bio-alcohol can be viewed as a synfuel produced from nuclear electricity and sunshine as prime energy sources. Under that scenario, non-portable nuclear energy is converted into portable energy and the 80% to 90% energy penalty for harvesting and extraction is not critical. Globally, assuming 40% of total energy consumption requires portable fuels, one would need a formidable 2.4 billion acres or 30% of the world's arable lands. Only in countries like the USA and Brazil with vast fertile land areas can such bio-fuel production be sustained, *provided sufficient nuclear electricity is also available*. Although bio-fuels like alcohol emit globe-warming carbon-dioxide upon combustion, there is no net addition of this gas to the environment since plants take up carbon-dioxide to make extractable carbon-based bio-fuels.

Solar cells converting sunshine directly into electricity, have been under subsidized development for more than 50 years. The best units have overall efficiencies of about 8% (including losses from voltage up-conversions and storage), yielding about 437 kW per acre of solar panels when sun-shine peaks. Allowance for diurnal, seasonal, and weather fluctuations reduces this value by 75% to an average 109 kW per acre, while a further reduction of 25% of land area for access roads to install and maintain solar panels, storage batteries, and transmission lines, yields a final 82 kW per acre. To replace all present petro-fuel and electric energy generation with solar-cell energy requires (9.1 billion kW)/(82 kW/acre) = 111 million acres of sunny open land. Construction of a large solar complex generating a million kW of electricity costs about $ 10,000 per kW and requires 12,500 acres of desert land. To provide 9.1 billion kW world-wide by solar cells then requires $ 91 trillion of capital investment and 111 million acres of accessible desert. Should all future homes possess solar panels to provide self-sufficient domestic power (30% of total energy pie), $30 trillion might be passed on from the utilities to home-owners, who probably will pay double this amount ($ 60 trillion). Utilities must still invest $ 61 trillion in this case.

The economics of windfarms is also fraught with location restrictions and large-area problems. Assuming these problems are solvable, wind-power is estimated to require an investment of $64 trillion to deliver 9.1 billion kW to the world. This assumes 1,250,000 advanced 2 MW(e) wind turbines together with power distribution systems and modern electric storage capacitors when winds are not blowing hard enough (80% of the time). In contrast, to supply the world's electricity needs and replace all petro-fuels with synfuels produced with nuclear heat or electricity from reactors that generate 1 million kW each, requires (9.1 billion kW)/(1 million kW) = 9,100 nuclear reactors world-wide. Real estate for 9,100 reactors comes to 364,000 acres, assuming each reactor takes 40 acres for buildings and cooling towers. They can be built anywhere away from earthquake faults. At $1.8 billion each, the 9,100 reactors would demand $ 16.4 trillion of capital. It is not difficult to guess what utilities and capital investors prefer when choosing between $ 91 trillion for solar, $ 64 trillion for wind, or $ 16 trillion for the nuclear option.

For the USA, replacement of primary oil and coal requires an investment of $35 trillion for solar, $25 trillion for wind, and $6 trillion for nuclear power.

Besides these capital cost disincentives, the enormous land areas needed for solar and wind energy cause a disturbance of local ecologies and will spoil many scenic landscapes. Exclusive use of these sources for prime energy would make them very unpopular with environmentalists. Aside from capital costs, one must consider maintenance costs. Solar cells require constant cleaning to remove dust or bird droppings, and must be replaced every ten to twenty years due to erosion and deterioration (sand storms, etc). They are made of gallium-arsenide or copper-indium-diselenide, requiring toxic silanes, arsenic, etc. for their manufacture. Toxic wastes generated in producing solar cells for global use, dwarf the amount of nuclear fuel and waste for the nuclear option. For wind-power generation, the mechanical maintenance of thousands of turbines and protective measures to avoid killing thousands of birds, seriously effects its economics. The secret of controlled nuclear power is that it is a thousand times more concentrated than any non-nuclear method.

3.4. A BRIEF HISTORY OF ENERGY

In 1650, the world was populated by 550 million people, or less than 10% of the present population. Besides sunshine which energizes agriculture, controllable energy resources available to man were:

Human Labor (via contracts, indenture, slavery, or prisoners to build structures, roads, etc).

Animal Labor (horses, donkeys, camels, elephants, dogs, for transporting people and goods).

Wood, Oils, and Coal (burned for lighting, cooking, heating, melting/forging copper and iron).

Wind (windmills grinding wheat, pumping water; sailing ships transporting goods and people).

Water Flow (waterwheels grinding wheat, aquaducts, drainages, etc).

These main forms of energy were all utilized in one way or another to satisfy man's basic needs and wants for water, food, warmth (heat), housing; or for manufacture of goods such as clothes, candles, furniture, saddles, carriages, ships, armor, weapons (for hunting, defense, and warfare), etc; or for moving people and goods (transportation). They are still available, but now we have 6 billion people.

In the 18th, 19th, and early 20th century, several discoveries and inventions were made that profoundly changed the world's energy picture. First came the steam engine, originally demonstrated by James Watt of Scotland in 1770. It burned coal that heated water in a boiler, converting it into steam which in turn pushed pistons that turned wheels. It was actually preceded by an 'atmospheric pump' that used condensing steam to pull a vacuum for suction, invented in 1712 by Englishman Thomas Newcomen, to pump water out of flooded mines. However it was not until 1807, after engineer Robert Fulton (USA) made improvements in the mechanical linkages and conversion cycle of heat to mechanical motion, that steamships and steam-locomotives were developed worldwide. Starting in the 1820's, steamships plowed the oceans and big rivers of the world, while trains pulled by steam-locomotives traveled over railroad networks all over the globe, connecting widely separated land-locked territories. Coal became a very important commodity and many new coal mines were opened to feed the hungry steam engines of the 1800's. Some coal-fueled steam-powered automobiles were also built, but only the rich could afford them.

Between 1878 and 1885, pioneers Elihu Thomson and Nikola Tesla learned to generate alternating current (AC) electricity in a continuous fashion, and subsequently invented the AC electric induction motor, electric lightbulbs, and many other products run by electric power. Because AC voltages are much easier converted and boosted, AC power distribution over copper wires is far less lossy than Thomas Edison's DC (= direct current) electricity that powered the street-lights of New York in the 1890's. AC electricity can be delivered fairly efficiently over large distances. George Westinghouse who supported and financed Tesla's work, built and exhibited the first commercial hydroelectric turbo-generator at the 1893 World Fair in Chicago and shortly thereafter completed the first large hydroelectric plant powered by waterfall-driven turbines at Niagara Falls in 1895. It delivered a 'whopping' 1.1 MW(e) of electric power that ran the lights and streetcars of Buffalo, NY, twenty-six miles away. General Electric built the first power lines for this venture. Invention of the coal-burning steam-driven turbine in 1884 by Sir Charles Parsons in England, provided an alternative to the waterfall-driven turbines used in hydroelectric power plants. In Parson's scheme, a boiler heated by burning coal converts water to pressurized steam which in turn drives a turbine that generates electricity. Steam turbines are now the major generators of electricity, and coal is the prime energy source for 52% of all electric power generation in the USA. Only 5% comes from hydro-electric sources, since most rivers in the USA suitable for dams and large-scale electric power generation have been exhausted. Additional steam-turbine electricity is presently generated by heat from uranium fission (21%), and by burning natural gas (12%), petroleum/oil (3%), or industrial waste (wood, biomass, alcohol) (5%). Geothermal steam, wind-power, and solar-cells provide the remaining 2%.

Two additional world-changing developments were the introduction of the petrol-fueled internal combustion engine for automobiles in 1889 made by Gottfried Daimler in Germany, and the flight of a petrol-engine-powered airplane in 1904 by Orville and Wilbur Wright. Mass-production of autos in 1908 by Henry Ford in the USA, and selling his cars on the installment plan, were another two revolutionary steps that changed the world. After 1908, mass-produced petrol-powered cars overtook steam automobiles and horse-drawn carriages, while most coal-burning locomotive and steamship engines were replaced by diesel engines. The rapid expansion of automobile usage and aviation, with increased demands for oil-derived petrol and diesel, created the large oil companies of today which recover, refine, and distribute enormous quantities of refined oil for a worldwide market.

In the Middle East, underground sources of oil were known to exist for centuries and exploited to provide fuel for oil-burning lamps and stoves. When it appeared that oil could power automobiles, William D'Arcey, an Australian businessman, obtained a sixty-year concession in 1901 to drill and extract oil from 500,000 square miles or five-sixths of what is now Iran [Ref. 12]. He formed the Anglo-Persian Oil Co, later to become Anglo-Iranian and still later British Petroleum (BP). Similarly, in 1904 in what is now Iraq, the Armenian C.S. Gulbenkian recognized the enormous potential of oil and persuaded the Turkish Sultan Abdul Hamid to transfer ownership of immense tracts of land from the Ministry of Mines to private ownership (mostly himself), establishing the Iraq Petroleum Co. Later on, BP and Royal Dutch Shell obtained contracts to exploit the oil fields in Iran, Iraq, and Arabia and to export the oil. U.S. companies Exxon and Mobil, which had their starts in the oil fields of Texas and California, entered the Middle East arena in 1928, when they became part owners of the Iraq Petroleum Company. Gulf, Standard Oil of California (Chevron), and Texaco got involved somewhat later. Control over the Middle East oil fields stayed firmly in the hands of these 'seven sisters' until 1973, when host governments demanded more control and revenue from their mineral wealth. Today the oil-producing nations of the world have united under OPEC (Organization of Petroleum Exporting Countries) which regulates the world's oil production rates and prices.

In summary, two energy sources already known in the middle ages but previously consumed in modest amounts, were suddenly catapulted into major world commodities, namely:

Coal – experiencing a large increase in demand after 1820 to empower steam engines; and after 1900 to vaporize water for electricity-generating steam turbines.

Oil – experiencing large-scale exploitation after 1901 to fuel automobile combustion engines.

Today's oil consumption continues to rise, driven by ever expanding fleets of petrol-burning transport vehicles and craft, while the continuously expanding use of coal is due to increasing demands for electricity. Oil has allowed mankind to transport goods and people locally as well as to any part of the world quickly at an affordable cost, while electric power has enabled man to develop numerous new manufacturing techniques, products, and services. Electricity is providing modern homes with light, heat, air conditioning, electric stoves, refrigerators, radios, televisions, telephone service, etc.

Increased use and availability of oil and coal has also promoted world population growth and a desire in the less developed countries to acquire modern comforts. Oil and coal consumption rates have thereby reached a level where depletion of oil is forecast to occur in a few decades (one generation), while coal is expected to last only one century if world demand continues to rise at the present rate.

Without oil, it is impossible to maintain current forms of transportation, and without trucks and airplanes it is not feasible to produce and distribute enough food to feed six billion people presently on our planet. Fortunately, nuclear fission was discovered in 1939, and sufficient extractable uranium and thorium has been found on earth to generate all the needed electricity for manufacturing portable fuels for the entire world for the next 1,500 years. It appears that divine intervention wanted man to discover this new energy source in time to avert a human catastrophe when oil runs out.

The history of the discovery of uranium fission is a very interesting one and worthy of elaboration. Most of the following information is from Richard Rhodes' fascinating book [Ref. 11]. In Italy in the mid-1930's, Enrico Fermi was bombarding uranium with newly discovered neutrons, and observed that neutrons (atomic mass $M = 1$) were absorbed by uranium (atomic mass $M = 238$), causing the latter to transmute into new product elements with different atomic mass. Based on previous research, he believed that the atomic masses of the new transmuted elements had to either gain 1 atomic mass unit or loose 1 or 4 units from the original atomic mass $M = 238$ of uranium[7]. But he found that some product atoms did not have any of the expected chemical properties for a species with say $M = 239$ and $Z = 92$. In Dahlem, Germany, at the Kaiser Wilhelm Institute (KWI), Otto Hahn, Lise Meitner, and student Fritz Strassman decided to redo Fermi's measurements. They also found products with $M = 144$ and $Z = 56$, disagreeing with prevailing theory that they should not differ by more than a few units from $M = 238$ and $Z = 92$.

While Hahn and Meitner were pondering this result, Hitler invaded Austria in 1938 and incorporated it into Germany. Lise Meitner who was Austrian, was faced with new ugly Nazi laws that suddenly applied to her. She learned that her government-funded contract with KWI was about to be canceled because she was part Jewish, in spite of pleas by her colleagues. With the help of a Dutch physicist (Dirk Coster) who picked her up in Dahlem, she fled Germany and went on a train to Holland. From there she went to Niels Bohr's institute in Copenhagen, Denmark for a brief rest before going on to Stockholm, Sweden to work with Karl Siegbahn. In December 1938, Meitner got a letter from Hahn telling her he had repeated the experiments with neutron bombardments of uranium, and after careful chemical analysis of the products, he and Strassman found one product was definitely the element barium with $Z = 56$, *not at all* close to $Z = 92$. Did she have any ideas how to explain that?

Meitner met with her nephew Otto Frisch during Christmas 1938 in Sweden, and discussed Hahn's letter with him. After expressing skepticism but still preoccupied with Hahn's observation of barium, Meitner suddenly remembered a statement by Niels Bohr that he believed the nucleus of an atom such as uranium was like a pulsating liquid drop. They then conceived of the possibility that a nucleus could split

---

[7] A nucleus is made up of Z protons and M - Z neutrons. Each of the Z protons have unit atomic mass and unit charge. They determine the total positive charge of a nucleus, hence Z is also called the atomic charge number. The mass of a neutron is almost the same as the mass of a proton, but it has no electric charge. The total number M of 'nucleons' is the sum of protons and neutrons in a nucleus and is called the atomic mass number of a nucleus. A given element has a fixed number of protons Z, but can have different 'isotopes' with different numbers of neutrons and thus different mass number M. The most abundant uranium isotope is U-238 with $M = 238$ and $Z = 92$, i.e. 92 protons and 146 neutrons, while fissionable U-235 has 92 protons and 143 neutrons.

in two halves during a drop-stretching waist-producing pulsation after absorption of a neutron. They estimated this could happen if the atomic charge number exceeded Z ≈ 100 because of repulsion between the two halves, each half being filled with many positively charged protons. Uranium with Z = 92 protons, was close to this value. The two 'fission' products should each have atomic charge numbers whose sum was close to 92. That is if barium with $Z_1 = 56$ was one product, the charge number of the other fission product had to be near $Z_2 = 36$ if protons were to be conserved. Fissioning uranium actually yields product atoms with Z_1 and Z_2 spread over a range of values rather than one particular set. But the proton sum $Z_1 + Z_2 \approx 92$ does hold. Additional calculations convinced Meitner and Frisch that their hypothesis was physically quite plausible. They also determined that the liberated energy had to be enormous: ~ 200 Mev per fission or 82 GJ per gm of U $-$ 235 $\approx$ 1 MW-day per gm U-235 ($\sim$ 40 tankfulls of petrol per gm U-235!).

Frisch who worked at Bohr's institute in Copenhagen returned to Denmark right after the Christmas 1938 visit with his aunt in Sweden. He told Niels Bohr what he and Lise Meitner had deduced from the data of Hahn and Strassman. Bohr himself had proposed the liquid-drop model for a nucleus, and immediately concurred with their conclusion, stating this was an important discovery. Next, Frisch put together an experiment using an ion chamber he had in his lab. On January 13, 1939 he found that the masses M of some products from neutron-bombarded uranium atoms detected by the chamber, were much larger than the usually observed protons (M = 1, Z = 1) or helium ions (M = 4, Z = 2). Indeed they had values of about half the mass of a uranium atom (M = 238). This was experimental proof that neutron-bombarded uranium can fission. Niels Bohr left by boat to lecture at Princeton, USA, where he informed his colleagues about the Meitner-Frisch findings on January 17, 1939, after receiving a telegram from Meitner. One of these colleagues was Enrico Fermi who had just arrived in New York on January 2, 1939 after returning from Stockholm where he had received the Nobel prize for his pioneering work with neutrons. Fermi had decided not to return to fascist Italy under Mussolini, because his wife was Jewish and faced persecution.

The splitting uranium story told by Bohr was quickly passed around by the small U.S.'nuclear club'. Within a week, experiments with ion chambers were conducted at the National Bureau of Standards, which confirmed the observations by Frisch. Shortly thereafter, Einstein wrote a letter to president Roosevelt warning him Hitler might develop a super weapon using uranium fission. Thence the secret US Manhattan Project was born. To prove that uranium fission could work on a macroscopic scale, Fermi designed and built the first nuclear reactor at the University of Chicago, which went 'critical' on December 2, 1942. Next, in an incredibly short two years, plutonium production reactors were erected at Hanford, Washington, and a gigantic gaseous diffusion plant was built at Oak Ridge, Tennessee, to separate fissionable U-235 isotopes from natural uranium (0.7% U-235, 99.3% U-238). At Los Alamos, New Mexico, a team of the brightest scientists in the world under the leadership of Robert Oppenheimer worked feverishly to develop a nuclear bomb, thinking Hitler might be ahead of them. After testing the first bomb at Alamogordo,

N.M. on July 16, 1945, WW-II was ended with the detonation of two additional nuclear weapons, one on August 6, 1945 at Hiroshima, and the other on August 9, 1945 at Nagasaki, Japan. The irony is, that the three fascist-ruled nations who had banded together to conquer the world, were ultimately defeated by an international group of superb scientists, many of whom, like Lise Meitner and Enrico Fermi, had been driven out of their countries because Hitler alleged they or their family members were ethnically inferior!

After WW-II ended in 1945, vigorous development of nuclear *power reactors* (not to be confused with weapons!) occurred worldwide for the purpose of generating electric power by converting fission heat → steam → electricity. Today uranium produces 21% of all electricity in the USA, 85% of all electric power in France, and close to 50% of all electric power in Japan. Other countries (e.g. China) are quickly following. Clearly the latest, and today's most valuable energy resource is:

Uranium (starting around 1950)

Aside from electricity, 'research' reactors can produce special 'radioisotopes' used in thousands of hospitals by physicians specialized in nuclear medicine. The radioisotopes are tagged onto special pharmaceutical agents used for diagnostics or therapeutic cancer-fighting applications. Radioisotope-tagged molecules are also used widely as tracers in biotechnology and pharmaceutical research, revealing biological processes and the effects of experimental drugs in the human body. Finally, about two-thirds of reactor heat which is rejected at the lower temperature in electricity-generating steam cycles (see footnote 3), can be used for desalinization of seawater (California) or for mass urban heating (Mongolia).

3.5. SUMMARY OF PRIMARY ENERGY SOURCES

Natural prime energy sources are either 'renewable' or 'non-renewable'. Non-renewables are extracted from the earth with energy expenditures that are a fraction ($\sim 20\%$ or less) of the potential heat of combustion of the energy source. However because there is only a finite supply, they are depletable, i.e. non-renewable. Renewable energy resources on the other hand are assumed to be always available. Thus one has:

(a) **Non-Renewable Sources:**

   (1) **Fossil Fuels: Oil, Natural Gas, Coal**
   (2) **Nuclear Fuels: Uranium, Deuterium**
   (3) **Geothermal Energy: Heat Pockets**

(b) **Renewable Sources:**

   (4) **Water Falls and Tidal Waves (Hydro Energy)**
   (5) **Sunshine (Solar Energy)**
   (6) **Wind Energy**

Of energy sources (1) through (6), only item (1), oil, natural gas, or coal, are portable and can be taken along in an automobile, truck, or airplane to power it. Refined oil yields portable petrol (hydrocarbon mixtures rich in octane (C_8H_{18})), and portable diesel (crude oil distillates with higher boiling point) which are liquid at room temperature. Natural gas (natgas) contains mostly methane (CH_4) but also fractions of ethane, propane, butane (C_2H_6, C_3H_8, C_4H_{10}). As mentioned, compressed at ~ 120 atm in portable high-pressure tanks, natgas has fueled car engines. Coal can of course be carried along and burnt to make steam that powers a steam engine as was done in the 1800's. Today most coal and natgas resources are burned in power plants to provide steam heat that generates electricity via a turbine.

Regarding item (2), nuclear powered ships and submarines have been built and nuclear rockets or aircraft are feasible, but it is not practical nor safe for automobiles to carry nuclear reactors under the hood. Nuclear fission power to propel aircraft and rockets has not been implemented because of problems arising in potential crashes. Electric-grid energy, whether produced by uranium, coal, or other means, is clearly not in a portable form that can be carried by surface vehicles or aircraft as a replacement for petrol, when oil and gas are gone. However electric-grid energy can be converted into portable synfuel energy, as discussed in the next section. In large cities and parts of Europe, electric trains have been developed that use (nuclear) electric energy from the power grid by means of sliding blades that contact high-voltage bare overhead wires or ground-level trenched conductors.

In the 1980's, geothermal steam ran a power plant near some geysers in Cobb, Northern California. But steam pressure was gradually lost and the plant was shut down after six years. Such reservoirs of steam are evidently exhaustible like oil. Better geothermal schemes are being explored in France, Japan, and Australia. In the Cooper Basin of Queensland, Australia, anomalously hot underground fractured rock formations of about one thousand km^3 have been found at a depth of 3 to 4 km. Water under pressure (~ 600 atm) is piped into these dry heat pockets or 'hot-beds' and pressurized hot water at $T \approx 250°C$ is extracted and returned to the earth surface where its heat is exchanged with a second loop that converts a low-boiling-point fluid into vapor which in turn drives a set of vapor turbines to generate electric power [Ref 47]. Water is continuously resupplied to the underground hot-bed at the same rate as it is removed. The scheme is similar to a nuclear Pressurized Water Reactor (Chapter 5) in which fissioning uranium heats circulating water pressurized at ~ 150 atm to $\sim 280°C$, whose heat is transferred to a secondary low-pressure water loop that produces steam for generating electricity. The Australian underground hot-beds are estimated to hold sufficient quantities of extractable heat (~ 160 millionGJ/km^3) to run thirty 1000 MW(e) electric power plants for 30 years. Eventually they will be exhausted, so we listed geothermal hot-bed energy as item (3) under non-renewable primary energy sources.

One might think that heat from the earth's interior core and mantle could replenish the heat removed from underground hot-beds, but thermal conduction through solid rock is too slow to allow this. Without thermal insulation by rocks, most of the earth's interior heat would have leaked out long ago. Occasional volcanic eruptions are the only outlets of heat from below the earth's mantle. Hot-beds are different. They are puddles of silicate melts which during the earth's creation came to rest within the earth crust at depths of 3 to 10 km from the surface. After the earth cooled, the silicate puddles became fractured rock but remained hot in areas where covering rock is highly insulating [Ref 47].

Renewable item (4), hydro-power, is derived from the kinetic energy stored in bodies of water (lakes and rivers) that move due to earth gravity forces. It is used in hydro-electric plants where the flow of river water or a water-fall is passed through hydro-electric turbines that generate electric power. There is no direct application of hydro-power to provide portable propulsion for land vehicles, except for river-craft that move downstream with the river flow. Today most rivers with suitable conditions to generate hydro-electric power have been dammed, and new river sources are essentially exhausted. Tidal wave energy is re-examined now and then but seems only marginally worthwhile to exploit.

The harvesting of renewable solar and wind energy is primarily useful in remote locations that need electric power. Assuming wind is available, wind turbines can provide 1 to 2 MW(e) peak power per turbine at an installation cost of $ 1 million/MW(e). Solar power may be useful in desert regions, but installation and maintenance costs of solar stations are still high even though prices of solar cell arrays have come down considerably in the last twenty years. Of course winds are not always blowing and the sun is not always shining, which limits the use of solar and wind energy in many parts of the world. To replace petro-fuels with solar and wind energy on a global scale was reviewed in Subchapter 3.3.

For transportation applications, wind-powered sailing ships have of course been used for millenia, while sail-driven or solar-cell-powered cars have been built and do exist. Today these are great for sports but they could not replace the combustion engine to run bulldozers, trucks, cars, or any mass transportation vehicles. One finds in general that solar and wind energy, while attractive, lack sufficient power density to compete with modern compact high-power engines and fuels, and with nuclear power generation. High-density power generation lowers capital equipment costs immensely compared to costs of large-foot-print systems that collect and concentrate solar energy for example.

Conversion of biomass or organic wastes into energy-carrying synfuels by fermentation, digestion, and other techniques is only economically feasible if the required processing energy is supplied by (nuclear) electricity or heat. As mentioned, in this case extraction of a biofuel (e.g. alcohol or fat) from biomass represents a synfuel whose inherent locked-up energy is made available for combustion due to the input of externally provided (nuclear) electric/heat energy for plant cultivation and biofuel extraction. Non-portable electricity or heat energy is thus essentially converted into portable fuel energy in this case. To replace all present usage of petrol with biofuels, it was shown in Subchapter 3.3 that besides the availability of (nuclear) electricity or heat, about one-third of all available land in the world would have to be used for farming bio-fuel producing plants. This is feasible only in countries with vast thinly populated fertile lands such as Brazil or the USA, but may be impossible for densely populated Western Europe. If one dictates that no coal or uranium can be used to generate primary electricity, and only bio-fuel production is allowed, one finds that one needs four times all arable land in the world to replace present petrofuel energy. Clearly this is an untenable proposition.

CHAPTER 4

TECHNOLOGIES FOR PROPELLING CARS, TRUCKS, TRAINS, SHIPS AND AIRCRAFT

Providing vehicle locomotion generally involves two separate but interdependent components: (1) a portable energy source in the form of a battery (stored electrical or flywheel energy) or a portable fuel (stored chemical energy); and (2) an engine or motor that consumes the portable stored energy and converts it into mechanical motion. Various portable fuels and energy storage devices have been developed in the last century, the champion fuel being petrol because of its low cost and availability, and the champion battery being the lead-acid device because of its ruggedness and rechargeability.

Engines can be divided into three catagories: (I) motors driven by the electricity from chemical or mechanical storage batteries, (II) internal combustion engines (ICEs) fed by portable chemical fuels that can react with oxygen in the air, and (III) fuel-cell engines (FCEs) fed by chemical fuels that react with atmospheric oxygen. Motors in catagory (I) convert electricity from a storage battery by induction into mechanical motion of the wheels of an automobile via a series of gear trains, with an efficiency of 85% to 95%. ICEs in catagory (II) convert the heat of fuel combustion into mechanical motion via piston action with an efficiency of 30% to 40%, while FCEs in catagory (III) utilize an electrochemical reaction of fuel with atmospheric oxygen on special electrodes with an efficiency between 45% and 85%, whose electricity drives a motor as in (I). These efficiencies are ratios of energy delivered for mechanical locomotion over energy provided by a battery or fuel. Although FCEs are more efficient than ICEs, because of problems discussed below, the ICE has won out in the automotive field which uses portable fuels. It is presently the most developed device for propelling cars. Motors driven by batteries also lost out against ICEs because of driving range limitations discussed below. In what follows, we first review portable fuels and energy carrying batteries, and then discuss ICEs and FCEs.

4.1. REVIEW OF PORTABLE FUELS AND OTHER ENERGY CARRIERS

Portable fuels and portable batteries that can power small craft can be grouped as follows:

PORTABLE ENERGY SOURCES

(a) **Natural Non-Renewable Fuels**

Oil (Liquid Hydrocarbons – C_mH_n)
Compressed Natgas (Methane, Ethane, Propane, Butane: CH_4, C_2H_6, C_3H_8, C_4H_{10})
Coal (C)

(b) **Manufactured Energy Carriers (Synfuels and Batteries):**
Hydrogen (H_2), from Water (H_2O) + Electricity or Heat.
Ammonia (NH_3) and Hydrazine (N_2H_4), from Water (H_2O) + Air (80% N_2) + Electricity or Heat
Syn-Petrols and Methane ($C_mH_n(O_k)$, CH_4), from Water (H_2O)+Coal(C) + Electricity or Heat
Methanol (CH_3OH), Ethanol (C_2H_5OH), from Biomass (Corn) + Sunshine + Electricity
Rechargeable Electric Storage Batteries
Rechargeable Mechanical Flywheel Batteries

As long as large quantities of energy cannot be transmitted wirelessly (Tesla's dream), there will always be a need for portable energy sources. At present, portable fuels used in land or sea transportation and in aircraft propulsion come mostly from the non-renewable resources listed under (a), that is: Oil (Petrol or Diesel), Natural Gas (Compressed in Cylinders), and Coal. When these non-renewable resources are depleted, artificially manufactured energy sources must be developed which can be carried on-board vehicles to fuel engines or energize motors that propel the vehicles. Such portable energy can come in the form of a portable (synthetic) fuel that can be refilled periodically into an on-board fuel tank, or as an energy storage battery which is periodically recharged. In what follows we briefly examine past and present activities to develop portable synthetic fuels and rechargeable energy storage devices.

4.1.1 Portable Synfuels

The only practical solution to the out-of-oil problem is to manufacture portable fuels in which power-plant heat or electricity is converted (with some loss) and stored as chemical energy in an oxidizable compound. Such portable synfuels are most economic if they can react chemically with atmospheric oxygen (O_2) to produce heat or electricity and thence propulsion. Preferably they are synthesized from air (80% N_2; 20% O_2) and water (H_2O), both of which are abundantly available on earth. Hydrogen (H_2) synfuel would be advantageous since it can be made from water and returned into water when consumed, without polluting the atmosphere. It can be burnt in internal combustion engines (ICE) producing heat and propulsion, or consumed by a fuel-cell engine (FCE) to propel a car. A problem with gaseous H_2 is its high storage volume, the main reason why it is not (yet) widely used as a clean fuel for ICEs to replace 'dirty' petrol.

Compact electrochemical fuel-cells have been developed that consume tank-supplied hydrogen (H_2) and air-supplied oxygen (O_2), which can make sufficient DC electricity at near-ambient temperatures to power a car (see Section 4.2.2). Electric motors to turn wheels with this electricity are well-developed today and hybrid fuel-cell-driven automobiles are now (2004) coming on the market. Fuel-cell engines that are run on pure hydrogen, exhaust only water (H_2O). No nitric oxides (NO_x) nor carbon dioxide (CO_2) are emitted because they operate at a much lower temperature than that of a combustion engine. Thus fuel-cells are ideal car engines that do not pollute the environment. Interests of environmentalists and future-world planners have coincided here to promote this power source for (future) autos and trucks. As discussed in the next chapter, there are still problems to be solved for fuel-cells. As mentioned, one major problem is the fact that hydrogen gas is difficult to store compactly. Nevertheless hydrogen (H_2) is a potential universal synfuel, produceable from water by electricity or heat (provided by nuclear, coal, or other prime energy source), which can power ICEs and FCEs without CO_2 air-pollution.

Besides hydrogen (H_2), other potential portable synthetic fuels which can be made from atmospheric nitrogen (N_2) and water (H_2O) as feeds, are hydrazine (N_2H_4) or ammonia (NH_3). Hydrazine is a liquid, and ammonia is liquid when compressed at a modest pressure of 16 atm. Both have chemical energy stored in them comparable to petrol, and both produce heat in a combustion engine or can generate electricity via a fuel-cell when reacted with oxygen (O_2) from the air. Hydrazine is presently used as a rocket fuel, while liquified ammonia is used as a fertilizer in agriculture. An experimental ammonia-burning ICE was succesfully tested some time ago in The Netherlands. If it is found difficult in the future to fly airplanes with pure hydrogen as fuel, a possible alternative would be to use hydrazine or ammonia as a fuel, if this can be done without significant air pollution. Before these synfuels are acceptable for ICEs, byproduct NO_x gases from high-temperature combustion of hydrazine or ammonia must be converted to O_2 and N_2 by catalytic conversion devices used in many of today's automobiles. If used in colder fuel-cells, NO_x exhausts from hydrazine and ammonia fuels are absent, since FCEs yield essentially only N_2 and H_2O as exhausts. The advantage of ammonia and hydrazine is that they are much easier to store than hydrogen. Ammonia is as stable as natgas, but hydrazine decomposes when slightly heated. The latter may be stabilized when hydrated and/or loosely bonded to a complex. Though pure hydrogen would be preferred for FCEs, ammonia and hydrazine can be catalytically decomposed to hydrogen and nitrogen on FCE electrodes by alloys of platinum, ruthenium, osmium, and iridium. Because of easier tank storage of ammonia, it may be preferred over hydrogen for automobile FCEs, even though the required catalytic decomposition of ammonia on FCE membranes is a complication that consumes a little extra energy.

It was reported recently that nitrogen (N_2) can be compressed into a solid compound of pure trivalent N that stores a considerable amount of chemical energy. If quasi-stable, it could be used in ICE devices when decomposed or detonated by an electric spark. The exhaust would be environmentally harmless N_2 which is already present in our atmosphere at 80%. However the first succesful fabrication of this pure N compound required 1,725 °C and 10^6 atm of pressure! While other ways may be found to produce it, it is doubtful that it is stable enough so it can be handled and used as a portable synfuel.

Synthetic portable liquid hydrocarbons can also be made from coal (C), water (H_2O), and electricity, yielding man-made petrol-like compounds ($C_m H_n$), as is done in South-Africa at its SASOL plant. However coal-derived synfuels, if massively consumed in combustion engines, would produce extra globe-warming carbon dioxide (CO_2) gas. Thus the use of SASOL-produced fuels in a pollution-intolerant society may be restricted to non-automobile applications such as plastics manufacture.

Extracting liquid alcohol (C_2H_5OH) and methanol (CH_3OH) from agricultured corn or other suitable plants is another scheme to make hydrocarbon synfuels. Growing corn requires sunshine and electric power for fertilizer production and farm operations. Thus non-portable (nuclear) power-grid electricity and solar energy is converted into a portable fuel. Since the same amount of CO_2 gas is returned to the biosphere in combustion as was taken in by plants during photosynthesis, there is no net globe-warming biosphere pollution (unlike SASOL). Such biofuels are thus legitimate non-polluting synfuels even though they emit carbon-dioxide and require one-third of all lands to grow enough for the total replacement of petro-fuels.

As discussed in Subchapter 3.3, unless (nuclear) electricity is available for agriculture operations and processing, there are unsurmountable problems if one considers replacing petrol *only* with bio-alcohol in an out-of-oil economy. Energy budgets for farming, fermenting, distilling, manufacture of farm equipment, etc, show that the input energy to obtain alcohol from corn or sugarcane may exceed the energy present in the final alcohol product. Farming improvements might lower the input/output energy ratio to 80% to 90%, but it means that in a no-oil, no-nuclear, no-coal economy, 80% to 90% of agriculturally grown alcohol must be put back into its own operations. Then to power the world's transportation fleets, one would need four times all arable land in the world for growing corn, which is of course impossible.

Hydrogen synfuel can be made by electrolysis of water using available grid electricity:

(4.1) $H_2O + \text{electricity} \rightarrow H_2 + \frac{1}{2}O_2$,

This scheme is reported to produce 7 cubic feet (15.36 g) of very pure H_2 at 2 – 2.25 Volts DC with an input of 1 kWh(e) [Ref 22]. Iron plates are used for the H_2 electrodes and nickel-plated steel for the oxygen electrodes. Assuming our standard energy equivalence (Section 3.1) that 10 kg of H_2 can produce 0.72 GJ(e) of electro-mechanical fuel-cell output energy, one finds that 1 kWh(e) = 0.0036 GJ(e) of electric grid energy can be converted into portable H_2 for fuel-cell electricity generation with an efficiency of $[1.536 \times 10^{-3} \times 0.72\,\text{GJ(e)}]/[3.6 \times 10^{-3}\,\text{GJ(e)}] = 0.307 = 30.7\%$. Assuming the grid electricity was generated from heat by a steam turbine at 33% efficiency, the overall conversion efficiency of making hydrogen fuel from latent uranium (or coal) energy is 10%.

Instead of electrolysis, hydrogen can also be produced using fission heat directly in a 'chemo-nuclear' reactor. This skips conversion of nuclear heat to electricity by a steam turbine. For example at 850 °C sulfuric acid (H_2SO_4) breaks apart, and with iodine (I_2) it can generate H_2 as follows [Ref 37]:

(4.2a) $H_2SO_4 + \text{heat}(850\,°C) \rightarrow \frac{1}{2}O_2 + SO_2 + H_2O$

(4.2b) $I_2 + SO_2 + 2H_2O \rightarrow 2HI + H_2SO_4$

(4.2c) $2HI + \text{heat}(450\,°C) \rightarrow I_2 + H_2$

or overall :

(4.2d) $H_2O + \text{heat} + \text{Catalysts} \rightarrow \frac{1}{2}O_2 + H_2$

This 'sulfur-iodine' catalyzed scheme using nuclear reactor heat is under investigation, and may be compared with the older 'steam-reforming' method of making H_2 (now undesirable because CO_2 is made):

(4.3) $2H_2O + C + \text{heat} \rightarrow 2H_2O(\text{steam}) + C \rightarrow CO_2 + 2H_2\,(T \sim 975\,°C)$

The efficiency of producing H_2 by (4.2) or (4.3) using (nuclear) heat has been estimated to be about 60%, meaning 1 GJ of (nuclear) heat can be converted to 0.6 GJ of H_2-carried chemical energy.

Before selecting H_2 production by either one of reactions (4.1), (4.2), or (4.3), it is prudent to consider the entire picture of a future 'hydrogen economy'. Some questions are: is it most economic to make hydrogen in large quantities at several large conversion plants and to distribute it through pipelines like today's natural gas for delivery to fueling stations; or is transport in the form of compressed or adsorbed hydrogen by tanker trucks better, as is done today with petrol? Or is it more efficient and practical to employ existing electric grid lines that already distribute power widely, to generate hydrogen by electrolyzing tap-water at city fuel stations equipped with electrolyzers. For fuel-cell usage, H_2 must be quite pure (Section 4.2.2). If contaminated by mercaptans for example, it must be passed through special scrubbers to remove them. Rather than piping H_2 to commercial fuel stations, hydrogen could also be distributed directly to homes using pipe-lines similar to natgas distributions. Then customers can directly fill their H_2 fuel tanks at home. The hydrogen might be adsorbed in an automobile fuel 'bladder' (fuel tank of the future). With a little heat, the bladder releases H_2 again to run the auto's fuel-cell. Alternatively, electrolysis of tap water (generating H_2), might be carried out in people's garages at night with grid electricity. Because 'slippery' hydrogen effuses through many materials, and embrittles or attacks a number of metals, current natgas pipelines are unsuitable for distributing hydrogen. Thus new (larger) pipelines would have to be installed for the ditribution of hydrogen. Another concern with home-provided or -generated H_2 fuel would be safety, similar (but more stringent) to present-day home delivery of natgas.

Without considering energy expenditures for hydrogen gas distribution, the 60% efficiency of producing hydrogen by (4.2) or (4.3) using nuclear heat means that 1 GJ of latent uranium energy is converted into 0.6 GJ of hydrogen-held chemical energy. This in turn is convertible to 0.33 GJ(e) of electro-mechanical energy in a H_2 fuel-cell at 55% efficiency. Since 1 GJ of nuclear heat can also be converted to 0.33 GJ(e) of electricity with steam turbines, portable H_2 produced via (4.2) or (4.3) followed by fuel-cell consumption gives the same overall conversion efficiency of 33%. It puts nuclear fission energy 'under the hood of an automobile'. In contrast, we found overall energy conversion (4.1) by electrolysis to be only 10% efficient. However additional costs for distributing H_2 from large chemo-nuclear reactor plants by new pipelines or tank-trucks must be added and compared with electric-grid electrolysis of H_2O in individual H_2-fueling stations to determine total H_2 fuel costs. As mentioned, the advantage of the latter scheme is that wide-spread distribution of electricity by power grids is already in place.

A possible list of foreseeable future synthetic fuels or 'synfuels' made with the input of nuclear heat or nuclear electricity may be summarized as follows:

Hydrogen (H_2), from Water (H_2O) + electricity or heat.

Ammonia (NH_3) and Hydrazine (N_2H_4), from Water (H_2O) + Nitrogen (N_2) + electricity or heat.

Ethanol (C_2H_5OH) and Methanol (CH_3OH), from Bioorganics (Corn) + sunshine + electricity.

Syn-petrol ($C_mH_n(O_k)$), methane (CH_4), acetylene (C_2H_4), from Water (H_2O) + Coal (C) + heat.

From the analysis given above, for each GJ of heat expended in synfuel manufacture, between 0.1 and 0.6 GJ can be stored as portable chemical energy for the generation of electro-mechanical motion by an ICE or FCE. When consumed, hydrogen, ammonia, hydrazine, and biomass synfuels return the same chemical species used in their manufacture back to the biosphere. Their reaction chemistry in ICEs or FCEs reverses the synthesis steps. However coal-derived synfuels add new globe-warming CO_2 to the atmosphere, while biofuels require enormous farm fields besides electricity, making ammonia and hydrogen most attractive.

To satisfy world demand for portable synfuels when oil is gone and to avoid global warming, one finds that only heat or electricity from uranium fission plants can economically produce the large quantities needed for extended periods (Chap 3). Nuclear fission of uranium is able to support all synfuel production in the world for more than one thousand years. Proposed 'renewable' solar and wind energy sources cannot provide sufficient capacity at a reasonable and affordable cost. Geothermal steam may be added to the world's supply of electric energy for a while. But it is only available at a few locations on earth in limited quantity, facing exhaustion after a few decades. Nuclear power plants on the other hand can be constructed anywhere and can provide heat energy at a thousand times higher concentrations for millennia.

TECHNOLOGIES FOR PROPELLING CARS, TRUCKS, TRAINS, SHIPS AND AIRCRAFT 57

In conclusion, most portable synfuels of the future will more than likely be synthesized by nuclear electricity or heat with air and water as feed. Some bio-alcohols may also be in the synfuel mix, while plastics and other hydrocarbon-based materials will be derived from coal using nuclear electricity.

Note finally that for a chemical rocket going into space (with no air), it is necessary to provide two chemical reactants to produce heat and thrust. That is, both an oxidizer (liquid oxygen or peroxide) and a fuel (liquid hydrogen, hydrazine, or petrochemical) must be carried by a chemical rocket to propel it. On the other hand fission-heated nuclear rockets only need hydrogen. Auxiliary power systems on space vehicles likewise cannot depend on atmospheric oxygen. Nuclear decay heat is therefore usually provided by a reactor-produced radioisotope; or solar cells are used if there is adequate exposure to the sun.

4.1.2 Electric Storage Batteries

While electric-grid energy cannot be conveniently carried along by a vehicle unless electricity-carrying guide-wires or rails contact it continuously as is done for trains, trams, and city buses equipped with contact blades, it is possible to deposit electric energy in portable storage batteries to power automobiles. Unfortunately energy/weight ratios of the lightest known storage batteries are rather low, making it impractical for them to compete with present automobile combustion engines that are fueled with petrol. Only for special situations like golf-carts, wheelchairs, and short-range urban vehicles with 80 km (50 mi) ranges, are electric storage batteries useful.

In 1995 the State of California demanded that the automobile industry produce some electric cars by 1998 to combat air pollution. So a number of electric cars powered by stacks of batteries were built. Batteries can store electric energy from the electric power grid and if they are rechargeable, they can serve as clean non-polluting portable energy carriers. The results over the past six years have not been promising however. The best purely electric models developed by GM, Chrysler, Nissan, and other car makers (Ford produced a hybrid later – see below) carry 150-500 kg of battery weight, and typically have a maximum range of 100 km (60 mi), while requiring 3 to 6 hrs of battery recharging time. Rechargeabilty features of existing (lead acid) batteries and their power and energy capacity per unit weight were improved somewhat, but performance of these electric cars is still far below that of a modern petrol-fueled automobile. Most electric cars are operated at 300 – 330 Volts, with outputs of 250 W/kg fully charged dropping to 100 W/kg at 80% discharged, and capacities of 32 Wh(e)/kg for standard lead-acid units. For newer lighter glass-fiber reinforced lead-acid batteries, these numbers were improved to 500 W/kg(full) → 240 W/kg(80%) and 42 Wh(e)/kg. Later models using nickel metal hydride (NiMH) and zinc-bromine-based batteries claim 75 Wh(e)/kg and possible ranges of 280 km (175 mi), while new lithium polymer and lithium-manganese dioxide systems ($LiMnO_2$, $LiMn_2O_4$) claim yields of 200 Wh(e)/kg and a potential 450 km (281 mi) range. Using our standard 0.7 GJ(e) = 200,000 Wh(e) to move a car 600 km (373 miles), the best of these batteries with 200 Wh(e)/kg, requires 1000 kg (2200 lbs) of weight for such a range!

The difficulty with storage batteries is that nature has a more or less fixed electron density per unit mass for chemical compounds so that useful available energies from electrochemical interactions are only on the order of 1 eV[8]. Combustion yields 10 to 100 times more energy per unit weight of fuel than what is retrievable from a compound in an electrochemical reaction, although only one third of combustion heat can be converted into electro-mechanical energy. Typically a tankful of petrol for a medium-sized automobile contains 60 liters (16 gallons), weighing 44.5 kg (98 lbs). Substituting this fuel weight with 1000 kg (2200 lbs) of batteries is clearly unattractive as long as there is petrol.

While electric storage batteries cannot compete with the internal combustion engine (ICE) in power per unit weight, in recently introduced 'hybrid' cars, a combination of both has been achieved (see Brief 3). For long-distance driving the ICE runs the car while in stop-and-go city driving, a 100 kg (220 lbs) electric storage battery moves it. The ICE recharges the battery when it is low. By our estimates the battery could take the car 60 km or 37 miles without recharging. A hybrid car still produces globe-warming CO_2 of course, but less than a conventional ICE automobile. Fuel cells also depend on electrochemistry, but have the advantage that oxidant (O_2) is supplied by air and waste product water (H_2O) is exhausted to air. Their fuels are light and thus they give better driving ranges than electric storage batteries can provide on one charge. Besides the absence of a fixed heavy storage battery load, they don't require frequent recharging. For this reason development of 'electric' cars are now focussed on fuel-cell engines (Section 4.2.2).

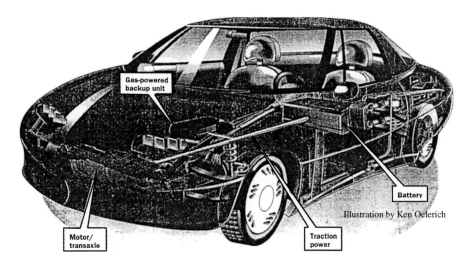

BRIEF 3. HYBRID ELECTRIC/ICE POWERED CAR

---

[8] The microscopic energy unit of eV (electron-volt) is used by scientists for energy exchanges between single atoms and molecules.

4.1.3 Flywheel Energy Storage (Mechanical Batteries)

Using induction coils, a possible portable energy storage scheme is to convert electric energy into mechanical energy of a high-speed rotating flywheel, and vice versa. This is sometimes referred to as a 'mechanical battery' and is used on some city buses (e.g. in Switzerland). It has also been considered for automobiles. However to store sufficient rotational energy for moving a car 600 km (373 miles) before a recharge, requires more than a thousand kilograms of flywheels. This is with assistance of fibers with the highest-known tensile-strength to hold spinning flywheels together.

An electro-dynamic flywheel device or 'mechanical battery' is a non-polluting energy storage unit which could be charged up in everyone's garage by electricity from the power grid. In such a direct electricity-to-electricity energy transfer between utility power and a portable non-chemical electric storage device, there is no need for intermediate conversions to chemical or electrochemical storage systems. In the flywheel scheme, counter-rotating pairs of magnetically suspended discs or tori are spinning at 180,000 rpm in evacuated cylinders as shown in Brief 4. They are electromagnetically coupled to induction coils that can extract electric energy or, in reverse, can store electric coil energy as mechanical energy of rotation into the spinning discs. As with electric storage batteries, one finds that engine weight is again a problem maker. The quantity of rotational energy stored in a flywheel is given by:

$$(4.4) \quad E = \tfrac{1}{2} \langle M \rangle (\Omega R)^2 = \tfrac{1}{2} \langle M \rangle V_a^2, \text{ Joules(e)}$$

Here Ω, V_a, and R are respectively the rotational velocity (radians/s), circumferential velocity (m/s), and outer radius (m), while $\langle M \rangle$ is the effective weighted mass (kg) of the rotating flywheel. For a torus-shaped (doughnut) flywheel with a thin body cross-section whose mass resides mostly at radius R, the mass $\langle M \rangle = M_{torus}$, while for a solid disc flywheel with distributed mass, $\langle M \rangle = \tfrac{1}{2} M_{disc}$. From (4.4) it is clear that the highest tolerable circumferential velocity gives the highest stored flywheel energy. An analysis of centrifugal forces, shows that the maximum tolerable value for V_a is:

$$(4.5) \quad V_m = (V_a)_{max} = (\Omega R)_{max} = (\sigma/\rho)^{1/2}, \text{ m/s}$$

Here σ is the tensile strength (Newton/m^2) and ρ is the density of the rotating material (kg/m^3). A fiber material which is believed to have the highest tensile strength is spider-silk. It was reported recently that spider-silk fiber molecules can be extracted from goat's milk. With a spider-silk fiber skin confining a heavy material, we estimate the maximum allowed V_m value is $V_m \approx 800$ m/s. If we assume our 'standard' 0.72 GJ(e) of energy is to be stored in a flywheel to achieve a 600 km (373 mile) range, one calculates from (4.4) with $E = 7 \times 10^8$ Joules(e) and $V_m = 800$ m/s, that $\langle M \rangle = M_{torus} = 2180$ kg (4818 lbs)! This mass could be distributed over 48 tori at 46 kg per torus, which are spinning inside 24 evacuated cylindrical chambers, each containing two counter-rotating tori.

Spinning at 180,000 rpm $\approx 10^4$ radians/s and with $V_m = 800$ m/s, one finds from (4.5) that the discs or tori must have $R = 0.08$ m $= 8$ cm or an outer diameter of

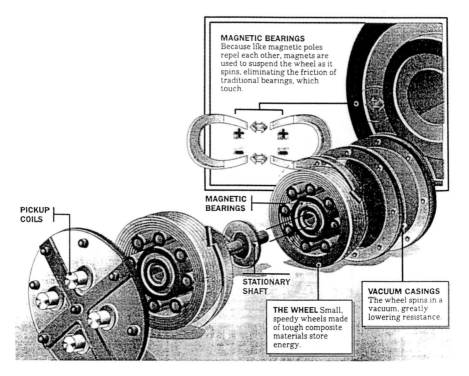

BRIEF 4. FLYWHEEL BATTERY COMPONENTS

$D = 2R = 16$ cm. Such an outer diameter is reasonable for cylinders that must fit under the hood of an automobile. This result is independent of torus or disc mass. A lower rpm will increase the value of D if V_m is fixed. With a torus of 8 cm outer radius and a torus body cross-section of S cm^2, the torus body volume equals $Q_{torus} = 2\pi RS = 50.27$ S cm^3. Thus if each torus mass is 46 kg = 46,000 gm, the density of the fiber-encapsulated material would have to be $\rho = 46,000/Q_{torus} = 915/S$ gm/cm^3. Using depleted uranium with $\rho = 18$ gm/cm^3, one finds that S must equal $S = 51$ cm^2 for each torus. This could be achieved if the uranium torus bodies had a circular cross-section with body diameter $D_b = 8$ cm or a rectangular cross-section with a width of 3 cm and height of 17 cm. Instead of 2180 kg (4818 lbs), a more reasonable total flywheel mass would be $\langle M \rangle = 400$ kg (881 lbs), giving a driving range of 107 km or 67 miles. Assuming again 48 tori, each now carrying 8.33 kg of depleted uranium, one finds $S = 9.2$ cm^2 in this case. This requires a torus body diameter of $D_b = 3.42$ cm if the cross-section is circular, or 2 cm × 4.6 cm for a rectangular cross-section. An 8.33 kg torus body is more convenient than a 46 kg torus body, but the driving range is of course only 107 km (67 mi) now.

Development of flywheel-powered automobile engines clearly depends on the availability of high-tensile-strength fiber materials such as spider-silk which can withstand high stresses at high velocities V_m to prevent flywheel disintegration. It is unlikely the weight problem can be overcome since it is improbable a super-material exists which can double the highest estimate of $V_m = 800$ m/s. Satisfactory solutions must also be found to minimize damage produced in auto crashes, in which stored rotational energy is suddenly released and converted into random kinetic energy. Using adjacent counter-rotating discs can partially neutralize such released energies. In addition, energy absorbing or deflecting skirts to minimize damage in case of an accident need to be explored and tested. Fires from a bursting gas/petrol tank in today's auto crashes can of course release as much energy in the form of heat, and are just as destructive.

A Chrysler-sponsored experimental flywheel-powered automobile at the 1994 Los Angeles auto show had a reported twenty cylinders under the hood, each cylinder containing two counter-rotating discs spinning at 200,000 rpm. It was reported to develop 100 kW(e) = 134 hp of mechanical power at start-up and to have a range of some 320 km (200 mi). By the estimates given above, such a range would require about 1000 kg of flywheel mass if $V_m = 800$ m/s. Instead of fully fly-wheel-powered cars, hybrid systems are being explored. Like electric hybrids, an ICE is used for long-distance driving and flywheels for short trips.

4.2. VEHICLE PROPULSION ENGINES

Besides well-developed electric motors empowered by electric or mechanical batteries, the three major prime movers which have been thoroughly investigated over the past one hundred years, and which require portable combustible fuels are:
1. **Internal Combustion Engines (ICEs)**
2. **Fuel Cell Engines (FCEs)**
3. **Steam Engines**

In what follows we shall briefly review these devices and examine their ability to consume future synfuels.

4.2.1 Internal Combustion Engines (ICEs)

The internal combustion engine (ICE) is the oldest device for propelling automobiles with portable petrol as energy source. It made its debut at the turn of the 20th century. Coal-burning steam engines had preceded the ICE during the 1800's, so the concept of moving a piston back and forth guided by a cylinder using heated vapor or steam was not new. However coal was burnt separately to evaporate water in a boiler while in ICEs, burning fuel transfers heat directly to expanding gas and is incorporated in it. Although others had suggested it earlier, N.A. Otto is credited with building the first succesful petrol-burning ICE in 1876, while G. Daimler first installed an ICE in an automobile in 1889 to succesfully drive it, using suitable transmission gears.

In the ICE, a mixture of air and petrol is admitted and compressed in a cylinder provided with a moving reciprocating piston connected to a crankshaft that turns wheels or gears. As the piston compresses the mixture in the closed cylinder, the latter is ignited

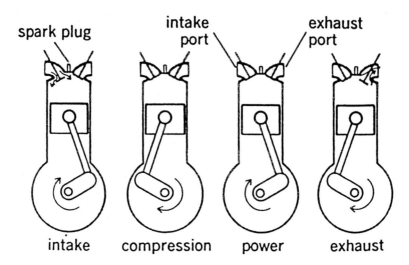

BRIEF 5. ILLUSTRATION OF INTERNAL COMBUSTION ENGINE FOUR-STROKE CYCLE

by a spark at the highest compression point just before the piston reverses its travel (see Brief 5). The heat of combustion then expands the gas which pushes the piston back in reverse which in turn rotates the crankshaft. After this work stroke, the expanded combustion gases are exhausted and a fresh mixture of air and petrol is injected to repeat the cycle. Many ingenious improvements to this basic concept were made during the 1900's, resulting in the sophisticated multi-cylinder automobile engines used today. The aircraft jet engine developed in the mid-1900's in England also burns petrol mixed with air in a confined space (internal), but instead of moving pistons, it utilizes hot gas expansion directly for propulsion.

Three main types of refined oil are produced to fuel present-day ICEs, namely aviation petrol, automobile petrol, and diesel. Petrol-burning ICEs use spark-plugs to ignite combustion, while diesel-burning engines self-ignite when compression temperatures and pressures reach a certain point. The hydrocarbon composition of diesel fuel is therefore different from that of petrol. Both type of petrols and diesel are obtained from crude oil but from different refinery distillate fractions.

Other means of locomotion will certainly be developed when oil reserves are depleted, but the ICE has been such a success during a whole century of development and usage, that it probably will not be totally abandoned. While today's ICEs run on petrol or diesel there is no reason other oxygen-burning synfuels cannot replace them to provide expansion heat. Alcohol (C_2H_5OH), ammonia (NH_3), hydrazine (N_2H_4), and pure hydrogen (H_2) all burn and react with oxygen (O_2) from air at certain ignition temperatures and mixing ratios. In fact alcohol and ammonia-burning ICEs have been built and tested, while hydrazine and hydrogen have fueled rocket engines for decades. Though DOT labels ammonia as non-flammable, ammonia/air mixtures can be ignited at 780 °C, burning with a yellow flame that yields nitrogen and water vapor. Mixtures of 16% – 25% ammonia gas in air can explode [Refs 24, 25] if ignited. Of course existing ICEs must be modified, since each synfuel has a different ignition point and requires different high-temperature containment materials. All hydrocarbon- and alcohol-fueled ICEs unavoidably produce unwelcome globe-warming carbon dioxide (CO_2) gas besides water vapor (H_2O), although bio-alcohols balance CO_2 exhausts with atmospheric CO_2 absorbed during plant growth. In addition, regardless of the fuel, high temperatures in ICE's produce some NO_x gases, since both nitrogen (N_2) and oxygen (O_2) reside in air intakes. For ammonia (NH_3) and hydrazine (N_2H_4) burnt in an ICE, NO_x byproducts may be somewhat higher. However nitrogen and water vapor can be the main exhausts if NO_x-removing catalytic converters developed for petrol-burning automobiles are used in future synfuel-burning ICEs.

As discussed in Subchapter 4.1, synfuels can be manufactured in large quantities using electric energy or heat supplied by nuclear and/or coal power plants. For the large numbers of automobiles used in ground tansportation all over the world, the use of non-polluting hydrogen-consuming fuel-cells appear to be the best solution to reduce future global warming when oil runs out and IC engines must be replaced. For aircraft propulsion where compact power systems are essential, it may be necessary and more efficient to use hydrazine-burning fuels and modified IC jet engines, provided NO_x byproduct exhausts can be eliminated. Future development programs and tests should determine whether fuel-cells can provide sufficiently compact take-off power to match that from today's large commercial IC-driven jet engines.

4.2.2 Electrochemical Fuel-Cell Engines (FCEs)

In a fuel-cell, chemical energy is directly converted into DC electric energy (see Brief 6(a)). In automobile applications, the DC voltage of about 1 Volt per cell is boosted by a stack of cells in series to about 120 – 330 Volts to drive four electric motors placed on each wheel or to run one motor and engage a gear-train that moves the car. Electrically driven automobiles have been under development for many decades and several well-tested schemes are available. The first publication of experiments with H_2/O_2 fuel-cells was by Sir William Grove in 1839 in England. Later, W.W. Jaques in the US and W. Ostwald in Germany, reported succesful direct electricity generation with high-temperature (500 °C) carbon/air fuel cells around 1896 and at the turn of the century. However interest in direct carbon/air fuel-cell generation of electricity waned, when more robust coal-burning steam turbines showed succesful conversion of heat to electricity at about the same time. Renewed efforts to produce electric power on a smaller scale with H_2/O_2 fuel-cells using modern materials, were started by Francis T. Bacon in 1933. Numerous other fuel-cell studies were undertaken after that by research teams all over the world. Several fuel-cell units with 2 to 20 kW(e) outputs were developed and used by NASA on Gemini, Apollo, and other spacecraft missions in the 1970's and thereafter. Briefs 6(b) through 6(d) show some early fuel-cell systems developed for NASA.

The fuel-cell is similar to an electric battery except the active agents are gases, usually air or oxygen (O_2) as 'oxidant' and hydrogen (H_2) as 'fuel', each being continuously supplied to one of two electrodes separated by an electrolyte or proton exchange membrane (PEM). Solid PEM's are preferred today over liquid (corrosive) electrolytes used in earlier devices. The fuel-cell process is inherently more efficient than a chemical reaction in a combustion engine where heat is produced first and then converted into mechanical motion. The latter conversion has a typical efficiency of about 33% set by thermodynamics which rules heat conversion. Fuel cells are 45% to 85% efficient in practice, losses being ruled by the kinetics in electrodes and electrolytes or PEM's.

Hydrogen and oxygen gases when mixed would like to react and form water by the overall reaction:

(4.6) $\quad H_2 + \frac{1}{2}O_2 \rightarrow H_2O$ (+heat or electric energy)

The reason is that the chemical binding energy of H_2O is higher than that of H_2 and O_2. At room temperature, a potential energy 'barrier' prevents them from undergoing reaction (4.6) however. There are two man-engineered pathways by which H_2 and O_2 can react to form H_2O. One is by the combustion process and the other via electrochemical action. The energy barrier in reaction (4.6) arises from the fact that the two H atoms in H_2 and the two O atoms in O_2 molecules are strongly bonded to each other. They must be taken apart first before they can recombine into H_2O. Because the total binding energies of H and O in H_2O are larger than the sum of the bond-breaking energies of H_2 and O_2, the difference manifests itself as kinetic energy, that is as heat in a gas for an ICE, or as electron motion in a FCE. In combustion, the heat increases the relative collision velocities of O_2 and H_2 molecules to the point that the bonds in H_2 and O_2 are broken, allowing for a re-arrangement of O and H atoms into H_2O. Kinetic energies required for prying O_2 and H_2 molecules apart into atoms and reforming them into H_2O molecules in a collision are on the order of 5 eV per molecule (see footnote 8). A gas mix of H_2 and O_2 will therefore not start to react measurably until an ignition temperature of about 580 °C is reached. Since heat is liberated after the reaction is initiated (e.g. by an electric spark), reaction (4.6) becomes self-sustaining in an ICE.

In the electrochemical scheme, electrolyte or a PEM is used to catalyze reaction (4.6) with liberation of electric energy. As shown in Brief 6(a), H_2 gas is bubbled over an electrode in contact with an electrolyte or PEM (proton exchange membrane; proton = ionized H atom = H^+) that helps dissolve or dissociate H_2 to become ionized with energy expenditures of about 1 eV/molecule (= 1.6021×10^{-19} Joule/molecule):

(4.7) $\quad H_2 + \text{Anode} \rightarrow \text{Anode}:H_2 \rightarrow 2H^+ + 2e^-$

Process (4.7) proceeds at much lower temperature (70–150 °C) than what is required in combustion. To get electrons to flow in an external circuit, O_2 is fed to a cathode placed on the other side of the electrolyte or PEM, where incoming H^+ ions are combined with oxygen atoms O and electrons:

(4.8a) $\quad \frac{1}{2}O_2 + \text{Cathode} \rightarrow \text{Cathode:O (adsorption)}$

(4.8b) $\quad \text{Cathode:O} + 2H^+ + 2e^- \rightarrow H_2O$

TECHNOLOGIES FOR PROPELLING CARS, TRUCKS, TRAINS, SHIPS AND AIRCRAFT 65

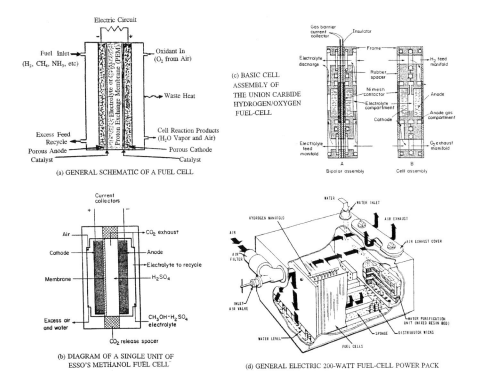

BRIEF 6. TYPICAL FUEL CELL SYSTEMS

Electrons liberated on the anode by (4.7) travel through the outside power-delivering circuit to the cathode to be reunited with H^+ ions, making neutral H_2O molecules via (4.8b) in accord with overall reaction (4.6). Protonization of H_2 by electrolytes or PEMs in (4.7) lowers the energy barrier in (4.6). If the freed electrons in (4.7) are allowed to flow through a conducting wire to be reunited with H^+ via reaction (4.8b) thereby helping to form H_2O, an electric current is promoted. Chemical energy is thus converted directly and efficiently into kinetic energy of electrons, i.e. electricity. A little heat helps increase the proton flow through the electrolyte or PEM and thereby increases the electric current. Presently, Dupont's Nafion is most popular for PEMs in experimental fuel-cells, but new high-temperature fluoropolymer materials to improve PEM performance are under development.

Hydrogen (H_2) is presently available as a cryogenic liquid or pressurized gas. Recent filament-wound high-pressure cylinders reportedly can hold 1 kg of H_2 each at 245 atm pressure. Each high-pressure fiber-glass cylinder weighs approximately 18 kg, has a storage volume of 50 liters, and nominal outside dimensions of 25 D × 125 L cm. For a 10 kg H_2 supply on board a car to travel 600 km = 373 miles (Brief 1), one thus needs ten cylinders. These cylinders could be mounted overhead or at floor-level in a car in two rows of five. They occupy about ten times the space of today's 60-liter petrol fuel tanks with dimensions of approximately 25 W × 40 H × 60 L cm.

An alternative solution for hydrogen storage are H_2-adsorbing porous 'bladder' materials of light weight that compact H_2 in a volume a thousand times less than in its gaseous state. The adsorption energy for such storage must be moderate so that H_2 can be expelled from the bladder again with little heat, without destroying or incapacitating it. Bladder materials under investigation are clathrates, (mixed) metal hydrides like MgH_2, $LaNi_5H_6$, $NaBH_4$, $NaAlH_4$:Ti, and (mixed) amides such as $LiNH_2$, LiN_aH_b, or $(LiH)_m : (LiNH_2)_n$. Various H_2-adsorbing carbon and boronitride nano-tube configurations are also under investigation. A capacity of 8 wt% (kg hydrogen per kg bladder) with an adsorption energy of 15 MJ/kgH_2 or less is the goal. If 10 kg H_2 is the desired quantity for one filling of a bladder fuel tank (Brief 1), such a fuel tank would weigh 125 kg without support structure. Presently the best materials still need improvements by a factor of four in weight as well as in adsorption performance to reach this goal. Note that hydrogen fuel storage costs extra energy, whether in bladder-heating to expel H_2 or in compressing/liquefying H_2 gas. An experimental fuel-cell car with ~100 kWe (~ 130 hp) of power has been reported to have traveled 400 km (250 mi) on a 155 kg (342 lbs) tank with 10 kg (22 lbs) of liquid H_2.

A second problem that has plagued fuel-cell systems is the fouling of electrodes and PEM's with particles and other 'poisons'. To minimize this difficulty requires careful filtering of the fuel. Additional schemes to cleanse electrodes and PEM's may involve intermittent ultrasonic exposures, intermittent 'flushing' with AC currents, or intermittent laser beam illuminations. This can be carried out at fueling stations or take place in a car owner's garage when the fuel-cell engine is not running (e.g. at night). PEM's could also be made replacable like the sparkplugs on IC engines.

Instead of pure hydrogen (H_2), liquid petrol (C_mH_n) or methanol (CH_3OH), and compressed methane (CH_4) or ammonia (NH_3) gas, can be used as fuel in automobile FCEs. These fuels are catalytically decomposed with liberation of H_2 on the proton exchange membrane. Alloys of platinum, ruthenium, osmium, and iridium seem to give the best catalytic performance. Although it adds some complexity and cost to the FCE, it may be a more practical and safer solution to the hydrogen storage problem than the use of compressed hydrogen gas at 240 atm pressure. Brief 7 shows an experimental car under development by Chrysler which uses a petrol-fed hydrogen-extracting fuel-cell engine. Such pioneers of the coming hydrogen age allow commercial development of fuel-cell technologies well before country-wide hydrogen (or ammonia) fueling stations and improved hydrogen storage techniques (or ammonia tanks) become available at a later date. Even though they still consume carbonaceous fuels and release CO_2, less air pollution (no NO_x) is generated by these engines than by ICEs, because of the lower operating temperature of fuel cells. This should please both environmentalists and futurists. Non-polluting pure hydrogen or ammonia would still be the most desirable fuel however after oil runs out.

TECHNOLOGIES FOR PROPELLING CARS, TRUCKS, TRAINS, SHIPS AND AIRCRAFT   67

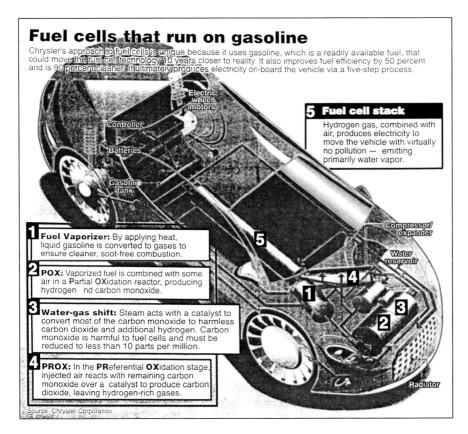

BRIEF 7. FUEL-CELL-POWERED HYBRID CAR

4.2.3 Steam Engine

For completeness, we mention one of the oldest automotive devices, namely the steam engine powered by portable coal as fuel. A return to using coal-burning steam-powered locomotives and automobiles of the 1800's has been proposed to counter the no-oil peril we face. Several experimental steam automobiles using modern components were built and tested in the 1970's, but further development was abandoned. Instead of burning dusty air-polluting coal, one could burn a liquid carbon-carrying synfuel (e.g. alcohol) to generate steam and propulsion. However in that case it is more efficient to burn the synfuel in an internal combustion engine than to use steam as an intermediary.

CHAPTER 5

ELECTRIC POWER GENERATION TECHNOLOGY

Historically the first electric power plants delivering large quantities of electric energy were hydro-electric. In 1895, Westinghouse built the first AC electric power generator in Niagara Falls by passing waterflow from the Falls through a turbine that induced electric currents in copper windings. General Electric built high-power grid lines that took the 1.1 MW(e) of electricity to Buffalo, New York, where it lighted the city's streetlights and powered streetcars. The technical brains behind this were Nikola Tesla who worked with Westinghouse and Elihu Thomson who was employed by General Electric (Section 3.4). Tesla who invented today's ubiquitous AC electric motor, was a Serb born in Croatia. He had studied electro-magnetics at universities in Austria, Tsjechoslovakia, and Hungary before coming to America in 1884.

Following the success in Niagara Falls, ambitious programs were started worldwide to dam up rivers and build hydro-electric power plants. Because suitable waterflow was not available everywhere, the alternative coal-fired steam turbine (invented by Parsons in 1884) was adapted for electricity generation: high-pressure steam instead of dropping water was used to turn a turbine's rotors which induce electric currents. Many coal-burning steam power plants were thus built in the early 1900's. Today (2004), 52% of all US electric power still comes from coal-fired power plants as illustrated in Brief 8.

After the discovery and demonstration of uranium fission in WW-II, heat from nuclear fission of uranium was used to replace the heat produced by burning coal. The same technique of generating high-pressure steam by heating water in a confined space is employed to drive the blades of an electricity generating steam turbine. The only difference between a coal and nuclear power plant is in the steam generation system and a coal-plant's air pollution. Everything beyond steam production is essentially the same for coal and uranium power plants. Since the central theme of this book is nuclear power, in what follows we focus on major aspects of nuclear electricity generation. Where appropriate, comparisons will be made between nuclear and coal or other power generating plants.

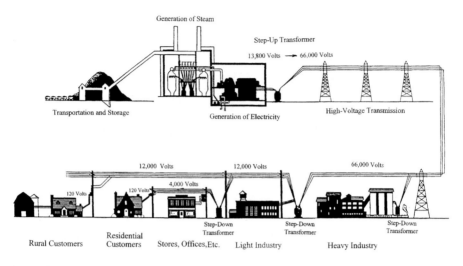

BRIEF 8. ELECTRICITY FROM COAL – FLOWSHEET

5.1. NUCLEAR POWER REACTORS

The first nuclear reactor was assembled in 1942 by Enrico Fermi and collaborators at the University of Chicago. Fermi who had left his native Italy for the US in 1939 when Mussolini's fascists took over, was one of a few pioneers in the 1930's who tried to understand the workings of atomic nuclei by bombarding them with newly discovered neutrons (see Section 3.4). After he learned in 1939 that his German colleagues Otto Hahn, Lise Meitner, and Fritz Strassman had shown that uranium could fission, he quickly started analyzing and measuring neutron multiplication by fissioning uranium and estimated what it would take to construct a 'critical pile' (now called a nuclear reactor). He determined the minimum amount of uranium and moderator needed to make neutron-induced fissioning of uranium self-sustaining. The first reactor was designed and built in 1941 according to Fermi's calculations. Using rods of natural uranium dispersed through a pile of graphite bricks, a self-sustaining chain reaction was achieved on December 2, 1942, with a 'critical mass' of uranium almost exactly as predicted by Fermi. Besides being a maverick in theoretical and experimental nuclear physics, Fermi had a social conscience. Like Leo Szilard and other colleagues, he disliked using nuclear fission for weapons. But he saw the urgency of developing a nuclear weapon in World War II before the nazis did, to defeat their tyranny and to defend the democracies.

Although the first reactor and several others that followed were not designed to generate electric power, the principle of operation of all nuclear reactors is the same. When the uranium-235 nucleus fissions, on average, 2.5 'fast' neutrons are liberated in addition to the two heavy fission fragments which carry most of the fission energy. As they slow down and stay in the solid material that encapsulates the uranium, the fission fragments heat it up. This heat is transferred by conduction to a coolant (water or gas) that passes through the core. In a power reactor, the heated coolant heats up water through a heat exchanger in a secondary loop and turns it into steam. The steam in turn drives a turbine that generates electric power.

Neutrons have the same mass as protons but have no electric charge so they can move through and bounce around in a solid material as if they were a gas. The lifetime of a neutron is approximately 12 minutes, which is essentially forever as far as the fission process is concerned. After a 'fast' neutron is slowed down (thermalized) by collisions in a good moderator material such as water or graphite, a thermalized neutron can be absorbed by a U-235 atom and cause it to fission, resulting in 2.5 new neutrons on average. After these are thermalized, the cycle is repeated and a chain reaction is started. While thermalizing, the gas-like neutrons 'fly' and diffuse through the solid core which is a mixture of moderator and uranium fuel. Then if neutron thermalization is sufficiently promoted by a moderator, a reactor can be made to run continuously in a self-sustaining mode. Because of so-called 'delayed' neutrons, reactors can be controlled safely at any desired level. The thermalizing neutrons provide some additional heat, but more than 90% of nuclear heating is delivered by fission fragments.

After Fermi's initial success, a number of 'research reactors' were constructed in the period 1945 through 1965 to further analyze and develop nuclear fission and its applications. Research reactors are still built and operated today to produce radioisotopes used in nuclear medicine for diagnostics and therapeutics (combating cancer), as well as to provide tracers for biotechnology research. In this case it is not the heat but the newly formed radioactive isotopes that are of interest.

The first uranium-fission power plant that generated electricity for a small community (Arco, Idaho) was operated in 1955 at the Idaho reactor testing station (now IN(E)L – Idaho National (Engineering) Laboratory). Many electric power reactors have been built since then and today 104 reactors provide power for 21% of the US electric grid. In France, 59 nuclear reactors provide 90% of its electricity needs. The total number of operating power reactors worldwide in 2004 was 438, with more under construction (mostly in Asia). In addition, the USA and Russia operate a hundred or so naval propulsion reactors.

The average age of the world's power reactors is about 25 years. Many are having their operating lifes extended for another 20 years after completing 30 years of their design life. Two reactor core meltdown accidents have occurred since 1955; one in Russia (Chernobyl) with a total of 45 fatalities, and one in the US (Three-Mile-Island or TMI) with no loss of human life. These accidents are discussed further in Section 6.3. The Chernobyl and TMI accidents caused a complete re-evaluation of reactor safety procedures (mostly human factors) and expansion of accident prevention techniques, originally believed to be adequate before these mishaps took place. Today, an accidental reactor core meltdown is as unlikely as a meteor hitting the Washington monument. Even if this happens, reactor containment vessels worldwide are designed to hold all nuclear debris. If a reactor overheats, the chain reaction is automatically stopped and the reactor shuts itself down.

5.1.1 Basic Design and Operation of a Reactor

The basic process in a nuclear reactor is the splitting of nuclei of uranium atoms by self-multiplying neutrons resulting in the liberation of heat. Physicists and chemists write this process in shorthand as:

$$(5.1) \quad ^{235}U + n \rightarrow FP1 + FP2 + 2.5\,n\,(+200\text{ MeV})$$

Here FP1 and FP2 are the two products from the split ^{235}U nucleus. They are new atoms with charge numbers $Z_1 + Z_2 = Z_U = 92$ (Z = protons in nucleus = 92 for uranium; see footnote in Section 3.4). There is a statistical distribution in possible product masses in uranium fission so we write FP1 and FP2 to describe them. A specific pair of products might for example be barium-144 with $Z_1 = 56$ for FP1 and krypton-89 with $Z_2 = 36$ for FP2. Fission products FP1 and FP2 cannot fission again but generate 'decay heat' while emitting betas (= fast electrons) or gammas (= energetic photons). They are lodged in the solid fuel elements and later removed as radioactive waste during spent fuel processing. Energy liberated in the fission reaction (5.1) is given in parentheses in atomic energy units of MeV (Million-electronVolt; 1 MeV = 1.6021×10^{-13} Joules). Most of the 200 MeV = 3.2×10^{-11} Joules per U-235 atom released in (5.1) is originally carried by FP1 and FP2 as kinetic energy, which converts to heat in the solid lattice which in turn is transferred to the reactor coolant. Since 1 gram of U-235 contains 2.56×10^{21} U-235 atoms, one finds that 1 gram uranium fuel provides 82 GJ = 22.8 MWh = 0.95 MWd of heat or 0.32 MW(e) = 320,000 W(e) per day, assuming the conversion efficiency of converting heat into electricity is 33%.

The 'core' in a reactor comprises twenty to a hundred or so fuel elements (the number depends on reactor power). Each element or assembly has bundles of fuel 'rods' which are zirconium tubes filled with uranium oxide pellets between which water can flow to remove liberated fission heat. The fuel elements are typically one to three meters (3 to 10 feet) long with square 10×10 cm (4×4 inch) cross-sections. Positioning pins at the top and bottom anchor them in so-called 'grid plates'. An array of such fuel elements then forms a roughly cylindrical core as illustrated in Brief 9.

Pressurized water at about 150 atm is made to flow through the core removing fission heat generated in the uranium oxide and acting also as neutron 'moderator'. The moderator decelerates all new neutrons originally born with very high kinetic energies in fission events. Moderators are desirable since neutron absorption by fissionable U-235 or Pu-239 is much enhanced for thermalized neutrons compared to that for high-energy or 'fast' neutrons. In a Pressurized Water Reactor (PWR), water under pressure is heated to about 300°C and passed from the core to a heat exchanger where its heat is transferred to a secondary low-pressure water loop. In this second loop, water becomes steam and is fed to a turbine. The expanding steam then generates electricity as shown in Brief 10. In BWR's (= Boiling Water Reactor), core-heated water/moderator is evaporated directly into high-pressure steam to drive a steam turbine, thereby avoiding a second loop. Both designs have pros and cons.

Instead of water, some reactors use gas as a coolant which drives a gas turbine instead of a steam turbine. Otherwise electricity generation is the same. If gas is used instead of water, moderator materials like graphite or beryllium are often used in fuel elements to effect neutron thermalization. Typical coolant gases are helium, nitrogen, and carbon dioxide. Still another reactor version developed in Canada is the heavy-water-moderated and -cooled CANDU reactor. This reactor uses D_2O instead of H_2O which has a lower neutron-robbing absorption cross-section. D_2O allows fissioning to occur with natural uranium (0.7% U-235; 99.3% U-238) instead of enriched (3% U-235) uranium needed in normal-water (H_2O) cooled reactors. To run a CANDU reactor, lots of deuterium (D) must be extracted from water (H_2O). Deuterium or heavy hydrogen (D = ^{2}H) occurs in concentrations of only 0.015% (D/H) in nature. A lot of energy is required to extract it from ordinary water (Canada has many hydroelectric plants and can do this with least expense). Today (2004) the world has 44 operating CANDU reactors, compared to 260 PWR's, 93 BWR's, 26 gas-cooled, 13 graphite-moderated light-water-cooled, and 2 fast-breeder reactors.

Surrounding the core of a reactor, there is usually additional water acting as a neutron reflector. The reflector in turn is surrounded by a 'gamma shield' that attenuates and stops most gammas emitted from the core. The entire reactor assembly is finally positioned in a containment vessel (Brief 9). As its name implies, the purpose of the containment vessel is to retain all fission products in the event of a core meltdown, strong earthquake, or other accident (e.g. an airplane crash on the reactor). The neutron reflector is very important and serves to minimize the loss of neutrons from the reactor and to maximize the neutron economy. Even though each

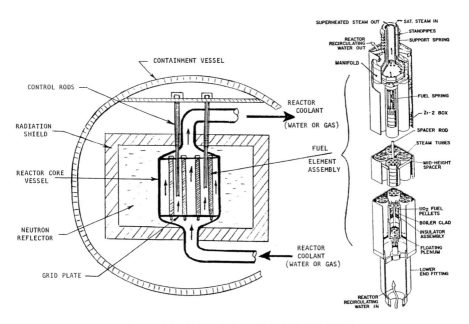

BRIEF 9. SCHEMATIC OF MAJOR NUCLEAR REACTOR COMPONENTS

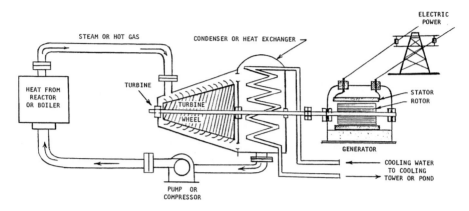

BRIEF 10. ELECTRIC POWER GENERATION BY A STEAM TURBINE

uranium fission produces 2.5 neutrons on average for each 1 that induces fission, there are losses due to neutron absorption by materials of construction (zirconium and steel), and due to escapes of neutrons from the reactor core region. The 1.5 neutrons that are left, while not needed to maintain the chain reaction, must be conserved as much as possible. In breeder reactors, absorption by uranium-238 of 1 of the 1.5 'left-over' neutrons is required to make new fuel. Neutron escapes are considerably reduced by a core-surrounding neutron reflector.

If too many neutrons escape from a reactor core, it is not possible to maintain a chain reaction. This can happen if not enough neutron slowing-down material (= moderator) is present and/or if there is no reflector. There is a minimum critical volume and associated critical mass below which a neutron chain reaction cannot be maintained because the surface area surrounding the volume is too large and the neutron escape too high. When a reactor is operating at steady state, the neutrons produced by fission and neutrons lost by absorption or escape are exactly balanced. A small increase in temperature will expand the surface area for neutron escape and make neutron moderation less effective to the point that the chain reaction is curtailed. Thus a nuclear reactor is inherently self-controlling and can only get overheated or melt down if a man removes the control rods (see below). Together with the 'delayed neutron' effect, this allows stable and safe reactor operations. Only with extreme operator negligence or deliberate sabotage in which control rods are forcibly removed, could a reactor meltdown be initiated when excessive generation of heat is not carried away fast enough by the coolant. Even this event is mitigated in today's reactors by sensors that in case of an abnormal excursion, activate controls to engage an independent emergency core-cooling system (ECCS), which in emergencies allows soluble neutron-absorbing compounds to be added to the coolant, thereby suppressing neutron multiplication.

To have a reactor operate at steady state with a constant rate of heat generation, the neutron population in a reactor is controlled by manipulating 'control rods'. These rods contain strongly neutron-absorbing materials like boron, samarium, europium, or other rare-earth containing compounds (Brief 11). Usually there are a minimum of four control rods which when completely inserted in the reactor core, make it impossible for the reactor to start or maintain a chain reaction. The control rods are pushed into the core with springs. To pull them out, electromagnets must first be energized after which a motor drive can slowly move them upwards out of the core to a desired preset level. If there is any disturbance, the currents in the electromagnets are interrupted and the springs push the rods back into the core in milliseconds. Current breaking is called a 'scram' and is usually initiated by the reactor operator. However in case excessive heat develops, or if cladding on a fuel element leaks, or if an earthquake strikes, the reactor scram is automatically activated by special sensors. Any one of such situations can cause an automatic non-operator-initiated scram.

In addition to control rods, some (research) reactors also have an emergency 'neutron-poison fuse'. This is a small rod with a high neutron absorbing material which is spring-injected into the core when a thin metal wire that keeps the fuse out of the core, melts and breaks under excessive heating. When this 'fuse' drops in the core, neutron multiplication stops and the reactor is sub-critical.

To start a reactor, one control rod is first slowly pulled out of the core by a motor activated with a switch under control of the reactor operator. As he pulls out the first rod, the operator watches a control panel to make sure no excessive neutron multiplication occurs. A small neutron source (often a polonium-beryllium source)

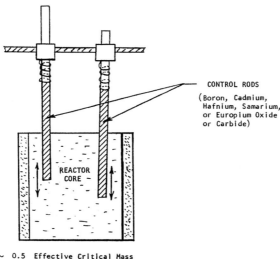

ALL RODS IN: ~ 0.5 Effective Critical Mass
ALL RODS OUT: ~ 1.25 Effective Critical Mass
RODS PARTIALLY IN: ~ 1.00 Effective Critical Mass for Steady-State Operation
KEY FEATURES OF REACTOR CONTROL PHYSICS: (a) Average Neutron Regeneration time;
(b) Delayed Neutron Fraction

BRIEF 11. ILLUSTRATION OF REACTOR CONTROL ROD OPERATIONS

is present in the core whose neutron emissions are detected by neutron counters (e.g. boron fluoride gas-filled tubes). As a control rod is pulled out of the core, the (sub-critical) multiplication of neutrons emitted by the independent source is measured and the amount of 'reactivity' or 'worth' is determined (reactivity is a measure of how close a reactor is to criticality). After the first rod is removed, the second rod is pulled out slowly next. Finally after the third rod is pulled up, the reactor becomes 'critical', that is it starts to increase the neutron population in the core by self-multiplication. The position of this rod is then adjusted and held when a desired amount of fission heat is generated steadily. A fourth rod called the 'shim rod' (and additional ones) stays in the core and is used for long-term corrections of the core's reactivity. With time, fissionable U-235 in the core slowly 'burns up' (requiring refueling every 1.5 to 2 years) so the reactor critical mass or 'reactivity' drops. To compensate for this reactivity loss, shim rods are gradually pulled out of the reactor core. We assumed four control rods here as in a small research reactor. Power reactors actually have banks of ten or more removable control rods distributed throughout the core to even out neutron flux distributions and thus fission heating, but the operating principle is the same.

Most nuclear plants that were built in the last two decades generate 1000–1200 MW(e) of electric power. Brief 12 showed the layout of a typical nuclear power plant cooled with gaseous carbon dioxide (CO_2), as designed and built in Great Brittain in the 1950's through 1970's. In general one has a reactor hall where reactor heat is produced and transferred to a coolant which in turn transfers its heat to water that turns into steam through heat exchangers in a secondary loop. The hot steam is passed onto the turbogenerator hall where it produces electricity by driving turbines. Low pressure ('exhausted') steam is condensed into liquid water in a condenser in the generator hall from where it is recycled back to the reactor room to pick up heat and be changed into steam again. The condenser which liquifies the exhausted steam, uses a second loop of cooling water (not the same as steam cycle water) which takes the final 'dump' or 'waste' heat from the condenser to a cooling tower. The cooling tower is usualy the most conspicuous feature of a (nuclear, coal, or natgas) power plant. It drops hot coolant water from the top of the tower downwards through large stacks of air-vented trays, thereby cooling the water through evaporative cooling. In some earlier power plants, water from a river was used to cool the condenser and the heated water was directly returned to the river without use of a cooling tower. With larger power plants some fish were found to die due to excessive temperatures in the river near the point where condenser exhaust water was returned to the river (labeled 'thermal pollution'). Today most power plants use cooling towers and/or a cooling pond before returning water to a river. Actually fish have been found to thrive and are attracted to lukewarm water from a power plant. As mentioned, steam-to-electricity conversion of a power plant is the same for nuclear and coal- or natgas-burning plants, and the above-described water-cooling operations apply to all.

Before government approval is given to build a nuclear power plant, a 'hazards analysis and environmental impact report' must be prepared and submitted by the

ELECTRIC POWER GENERATION TECHNOLOGY

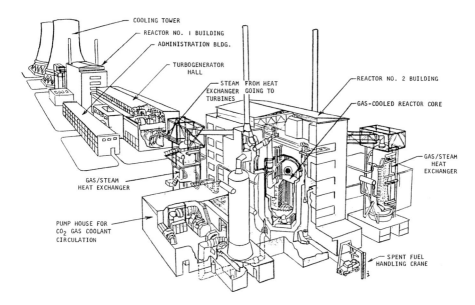

BRIEF 12. TYPICAL LAYOUT OF A NUCLEAR POWER PLANT

reactor construction company to the Nuclear Regulatory Commission (NRC). This report is carefully examined by the NRC (which usually takes over a year) to insure all required safety features are incorporated in the reactor design before it is approved. After construction, NRC representatives inspect the reactor for compliance.

5.1.2 Breeder Reactors

Because only 0.7% of natural uranium is directly fissionable U-235, and 99.3% is the less fissionable U-238 isotope, large amounts of uranium would be wasted unless the U-238 is converted to fissionable Pu-239 by the breeding reaction U-238 + n → Pu-239 + 2β. This breeding reaction takes place whether intended or not in every power reactor that uses say 3 percent enriched U-235 (reactor-grade) uranium, since the remaining 97 percent U-238 is always present to absorb neutrons.

Uranium reserves are not unlimited and would be exhausted in some 30 to 50 years if only U-235 with a natural abundance of 0.7 %, were burned up (see Brief 2). By converting U-238 to Pu-239 however, nuclear uranium reserves can provide energy for a period that is 144 times longer than what would be available from burning U-235 alone since the ratio U-238/U-235 = 144. Because of losses, the actual factor is 60, so that uranium reserves can provide nuclear energy for 1800 to 3000 years if breeder reactors are utilized. Clearly, operating breeder reactors are critical to long-term nuclear energy programs [Ref. 32].

The optimum situation for a breeder-based fission energy economy occurs if one can breed one fissionable Pu-239 atom from U-238 for each atom that is fissioned, that is a breeding ratio of 1 or more. This means that of the 2.5 neutrons produced in a U-235 or Pu-239 fission event, 1 neutron must be absorbed by a U-238 atom for breeding. Since an additional neutron is required to maintain the fission chain reaction, one can afford to loose only 0.5 neutron per fission for core escapes or absorptions by construction materials and coolant. To design a nuclear reactor such that only 0.5 neutron per fission is lost requires materials of construction and coolants with extremely low neutron absorption cross-sections. For example ordinary water (H_2O) cannot be used as coolant because it contains ordinary hydrogen (H). While H is the best (= lightest) moderator material, it also absorbs neutrons just a little too strongly. One must instead use liquid sodium, bismuth, or heavy water (D_2O) as a coolant and the uranium fissioning process must be shifted from a thermal (using thermalized neutrons) to an 'epithermal' or 'fast' fission process, employing epithermal or fast neutrons. Although fast fission lacks the convenient time delays of slowed-down thermal neutrons, there are means (fast neutron reflectors, etc.) to make operation of a fast-fission breeder reactor as safe as that of thermal-fission 'burners' (a 'burner' reactor has a breeding ratio less than one). In summary, control techniques, construction materials, and primary coolants are chosen differently for breeders than for burners, but in the main the same operating principles apply [Ref 32].

Most power reactors that have been built to date are thermal U-235 burners since they are less expensive to build an operate (by a factor of 1.1). Only three breeder reactors are presently (2004) operational, one in France, one in Japan, and one in Russia, each generating about 300 MW(e) of electric power. India has a breeder under construction. The USA once was in the lead and had a breeder reactor development program. However previous government administrations unwisely mothballed it, looking only at the short term. In the next ten years, it is imperative to phase in more breeders and less burners lest we leave our children with electric energy shortages in mid-century. It seems that France, Japan, and Russia have taken the lead and are acquiring important long-term operational experience with breeders.

The comments about thermal nuclear reactor operations apply equally to epithermal and fast breeder reactors. The main difference between breeders and burners is that plutonium must be (re-)processed and purified. This adds extra costs to the fuel cycle when compared with present burner operations that use only 'once-through' enriched uranium and discard plutonium- and U-238-rich burnt-up fuel as 'waste'. The plus side of a nuclear breeder economy is of course a millennium of energy availability. Also most generated plutonium will be continuously burnt up and does not accumulate, so the threat for possible weapons diversions is lessened. Strict accounting of all plutonium (like banks do with money) is necessary of course but quite feasible. The world has no choice if it wants to avoid serious shortages of energy by the middle of this century. Governments should assist electric power industries with development and financing of the essential plutonium reprocessing

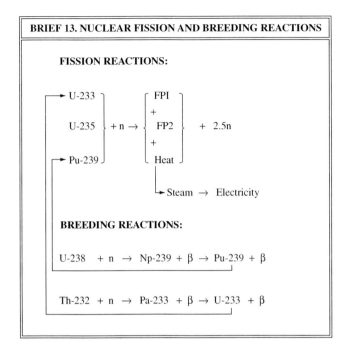

technologies, as well as with security against plutonium misuse. Because plutonium handling adds extra costs and breeder reactors are newer, utilities have been slow to finance and introduce them on their own.

Besides breeding U-238 to make fissionable plutonium, there is one other element, thorium-232, that can be bred to make fissionable U-233 by neutron absorption: Th-232 + n → U-233 + 2β. Reactors using thermally bred U-233 have been built and operated. This breeder cycle can thus be added to future U-238-breeding schemes when they become important. Estimated reserves of thorium in India, Brazil, and elsewhere, indicate that breeders using thorium can provide an extra 300 years of electric energy to the world. Brief 13 summarizes all possible nuclear fission and breeding reactions that can support neutron chain reactions, using available uranium and thorium resources from our planet.

5.1.3   Nuclear Reactors versus Nuclear Bombs

There is a mistaken popular belief that a nuclear power reactor is nothing but a controlled nuclear bomb. Nothing is farther from the truth. Although both make use of the nuclear fission process, there are major differences in their construction and modus operandi. As illustrated in Brief 14, a nuclear reactor with about 1.2 critical masses of uranium or plutonium fuel has its excess critical mass (0.2) nullified by the presence of neutron-absorbing control rods. On the other hand a nuclear

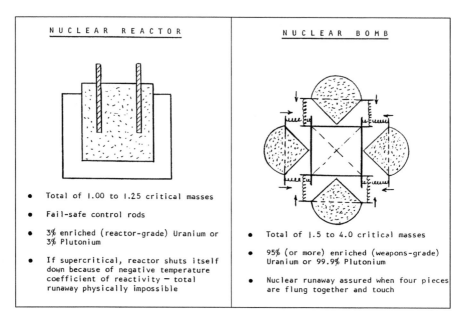

BRIEF 14. DIFFERENCES BETWEEN A NUCLEAR REACTOR AND A NUCLEAR BOMB

bomb has no control rods and between 2 and 4 critical masses of pure nuclear fuel (no moderator), which are kept apart before it is detonated. When flung together, these mass components make a super-critical assembly, causing an uncontrolled exponentially increasing multiplication of neutrons and fissions. The sudden release of an immense amount of heat then results in a fireball and explosion. Brief 14 illustrates the difference between a nuclear power reactor and a nuclear bomb.

Steadily running nuclear reactors, whether thermal or epithermal, employ thermalizing neutrons that require a certain 'slowing-down' time, which facilitates reactor control. Even in fast breeder reactors, fast-neutron reflectors impart delays in neutron regeneration lifetimes, which aids reactor control implementation. In addition there is a 'delayed neutron emission' phenomenon that assists in the control of all reactors (thermal, epithermal, and fast). The design of electromechanical control rod drives is such that rods can only be withdrawn slowly from the reactor core and only one after the other. This prevents sudden excursions and allows alarms to be set off in time to interfere with undesirable rod movements and/or to induce a scram. Even in the event someone would pull all control rods fully out of the core in an act of sabotage, the initial excess heat generated by such a reactivity increase causes a decrease in the coolant/moderator inventory because of the coolant's decrease in density with increasing temperature. Less moderator means less neutron slow-downs making the reactor subcritical. In addition, thermal expansion of the core and reflector increases the external surface area and thus neutron escape. The result is that the reactor ceases to continue the neutron chain reaction. This effect is referred to as a 'negative (temperature) coefficient of reactivity'. All modern gas- or water-cooled and -moderated power reactors exhibit this feature.

After shutting itself down, a reactor core is sub-critical but can still produce decay heat for a while which is removed by continued coolant flow. Should this normal flow be interrupted, nearly all reactors have an 'emergency core-cooling system' (ECCS) that kicks in to cool the core down. If both normal and ECCS coolant flows fail to perform, the core can still get overheated without uranium fission and melt itself down due to after-heat from nuclear decay. While some steam will be generated and some core components may melt or evaporate, the containment vessel, a thick steel pressure vessel that surrounds the reactor, is designed to retain all vapors and molten materials if this happens. No explosion in the sense of a bomb blast can ever occur, even in a meltdown.

Some time ago, alarmists in the news media reported that any freshman college student could design and build a nuclear bomb. These stories are difficult to reconcile with the fact that countries who actually have built and tested a nuclear weapon (USA, Russia, Great Brittain, France, China, India, Pakistan, Israel, South-Africa) spent three or more years and billions of dollars to get to that point. Sketching out a principle or a design on a piece of paper is not quite the same as building, procuring, and constructing equipment necessary for testing or use of a nuclear weapon. The same can be said about building a rocket to fly to the moon, which any college student can outline but not finance.

The uranium fuel used in nuclear power reactors usually has a U-235 enrichment of about three percent, whereas for nuclear bombs one needs U-235 enrichments of 95 percent or better if one wants an effective weapon. Thus there is a big difference in reactor-grade fissionable fuel and weapons-grade material. To increase the enrichment of reactor-grade fuel to weapons-grade fuel would require a very expensive uranium enrichment plant which is discussed in Section 5.2.1.

It is fortunate that building nuclear weapons is enormously expensive and difficult to carry out for a small group of terrorists. Even if they stole weapons-grade Uranium-235 or plutonium from an enrichment or reprocessing plant (Subchapter 5.2), it would cost a terrorist gang hundreds of millions of dollars of sophisticated equipment and a sizable work force that included some highly educated scientists to construct a usable nuclear bomb. Such an activity could hardly escape the attention of an effective global intelligence agency. It would be simpler for terrorists to steal an already manufactured nuclear weapon (still very difficult – Chapter 7) or use chemical or biological weapons to achieve their nefarious goals.

5.2. THE URANIUM FUEL CYCLE AND ENVIRONMENTAL IMPACTS

Uranium mining, enrichment, fuel fabrication, and radiowaste disposal make up the elements of the so-called 'Uranium Fuel Cycle', which we shall briefly examine in what follows. Briefs 15 and 16 illustrate the sequence of processes that uranium must undergo before it is ready as reactor fuel. Since the beginning of the atomic age, great care was taken to safeguard and assure uranium accountability. Concomitantly (with the exception of some early WW-II military operations), environmental caution in uranium mining and refining was stressed and strict safety measures were applied in the handling of uranium. The nuclear environmental safety disciplines that were instituted have always exceeded those in other industries such as coal mining and oil refining. Thus security and environmental awareness were promoted early because of the dual nature of uranium's application: weapons and (controllable) power.

Uranium ore has been found in may parts of the world such as Colorado, USA, Canada's Northwest Territories, South-Africa, Congo, Russia, and Australia. After digging it out of the earth (often on the surface), uranium-bearing ore is crushed and leached in a 'beneficiation' process usually close to the mine, which as final product yields a mixture of uranium oxides called 'yellow cake' with chemical composition U_3O_8. The U_3O_8 yellow cake contains natural uranium with 99.3% of U-238 and 0.7% of U-235 isotopes.

As mentioned, for a natural-water cooled nuclear reactor, it is necessary to increase the U-235 isotope concentration in uranium from 0.7% to about 3%. Some research reactors that produce special medical isotopes require high neutron fluxes so higher concentrations of U-235 are necessary, while for nuclear weapons a concentration of 95% U-235 or better is required. Special operations are therefore necessary to increase the natural concentration of fissionable U-235 from 0.7% to higher levels. This operation, called uranium 'enrichment' will be discussed in more detail later. The uranium enrichment process usually requires gaseous uranium hexafluoride (UF_6) as feed which must first be prepared from the U_3O_8 yellow cake by chemical conversion reactions. Thus as shown in Briefs 15 and 16, prior to entering an enrichment plant, U_3O_8 yellow cake is fed to a chemical conversion plant which produces UF_6. The UF_6 is then shipped to the enrichment plant in special autoclaves as a sublimable solid.

After the U-235 isotope in UF_6 is enriched, it is sent from the enrichment facility to a fuel element fabricator which converts it back to solid UO_2 pellets or metallic uranium. Fuel elements incorporate the enriched uranium as oxide pellets that fill zirconium tubes or use metal-clad uranium plates. One fuel element is generally composed of a bundle of tubes or a number of plates between which water can flow to carry off fission heat and to perform its function as a neutron thermalizing moderator. Fuel elements are typically 2 to 4 meter (6 to 12 feet) long with a 10×10 to 20×20 cm square cross-section. After shipment to a reactor, fuel elements are inserted and anchored to top and bottom 'grid plates' in a cylinder-like array which constitutes the reactor core.

Following two years or so of operation, fissionable U-235 in a fuel element is exhausted and the fuel element is removed from the reactor core and hung in a swimming-pool for a month or so to allow dissipation of initially high levels of decay heat and gamma radiation. The spent fuel elements are subsequently shipped to a reprocessing plant in a heavily shielded crash-proof casket. The shipment by truck or railroad can be carried out in an entirely safe manner and constitutes less

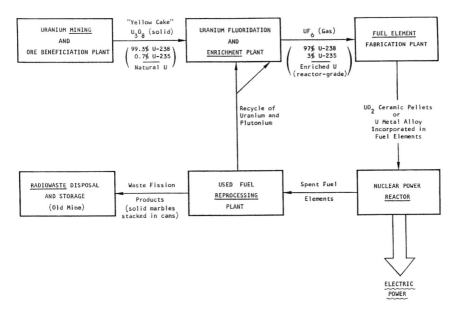

BRIEF 15. SCHEMATIC OF THE URANIUM FUEL CYCLE

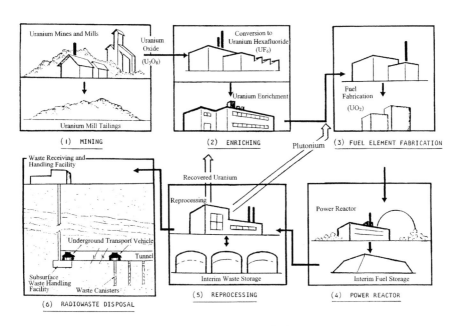

BRIEF 16. ILLUSTRATION OF THE URANIUM FUEL CYCLE

of a hazard than the routine shipment of some hazardous chemicals. Although the caskets are built to stay intact in a train crash, even in case they are penetrated by an armor-piercing mortar, little radioactive material could escape since the fission products are embedded in solid material. A statistical analysis for nuclear waste shipments from all one hundred US nuclear reactors, predicts two deaths per century attributable to radioactivity in worst-case casket transport accidents. This risk is fifty thousand times less than the estimated deaths caused by coal transports. Radioactive fission products cannot undergo further fission (only enriched uranium can). They only produce some heat and gamma radiation from slow nuclear decay.

In a used-fuel reprocessing plant, the canning or cladding on the fuel is removed and the fuel is dissolved in nitric acid. The uranium and plutonium fractions are then separated from the fission product fractions by chemical separation techniques and recycled (see Briefs 15 and 16). Some special fission products are also removed for use in radiomedicine and for isotope power generators used on space vehicles. The remaining fission products labeled 'radiowaste', are concentrated and vitrified into solid marbles which are placed in waste repository canisters. These canisters are shipped to a nuclear waste repository with underground vaults such as Yucca mountain. We now review uranium enrichment, reprocessing, and radiowaste disposal operations in more detail.

5.2.1   Uranium Enrichment

Although ordinary hydrogen (H) present in water (H_2O) is the best (= lightest) thermalizing agent to slow down neutrons, it unfortunately also absorbs a neutron every now and then instead of decelerating it. This happens to such an extent that a reactor using ordinary H_2O as moderator (= neutron thermalizer) as well as a coolant, must employ uranium that is enriched from 0.7% (natural) to 3% in U-235 to allow neutron chain reactions. Only with the next best thermalizer deuterium (D = ^2H) which is present in heavy water (D_2O), is it possible to use natural uranium with 0.7% U-235 to operate a reactor with D_2O as moderator and coolant. Very pure graphite (C) as moderator and helium gas as coolant can also operate a reactor with natural 0.7% uranium. Since ordinary water is universally available and abundant, most power reactors today use 3% U-235 enriched uranium ('reactor-grade' uranium) obtained from the enrichment plants built during WW-II.

As mentioned, for nuclear weapons one needs 'weapons-grade' 95% U-235-enriched (or higher) uranium, so during WW-II when the United States feared that Germany might develop a nuclear fission weapon before it did, a crash program was launched to build a uranium isotope separation plant of a magnitude and cost unprecedented in history. Under the command of army general Leslie Groves, a gaseous diffusion enrichment plant and an electromagnetic separation plant were designed and built in parallel, which were almost one hundred times larger and more costly than any other chemical plant ever built before anywhere in the world, and for which only scanty laboratory process data were available. Because one was not sure which process (electromagnetic or diffusion) would give the quickest results, one plant of each kind was designed and constructed simultaneously. There was less than one week's worth of natural uranium feed available for processing when construction of these two plants was started and many components were invented and researched along the way.

The electromagnetic separation plant at Oak Ridge, Tennessee was put together between February 1943 and November 1943 and immediately started enriching small quantities of uranium. It produced kilograms of highly enriched uranium used in the first two US atomic bombs (the third and last WW-II bomb used plutonium extracted from graphite-moderated reactors hurriedly built at Hanford after Fermi's success with the reactor pile in Chicago in December 1942). The Oak Ridge diffusion (DIF) plant took a little over two years to complete but was found more economic for large-scale enrichments than the high-vacuum electromagnetic 'calutron' method. In normal (non-war) times it would take ten years to design and build these two enrichment plants. After WW-II ended, the first power reactors obtained their 3% enriched uranium fuel from the diffusion plant, while the calutrons were (and still are) diverted to separate small quantities of medical and research isotopes.

The WW-II construction of the Oak Ridge uranium enrichment plants in Tennessee combined American ingenuity, organizational skill, brains, and a unity of purpose which will probably never be equalled again in US history. It remains as a unique once-in-a-thousand-years engineering achievement. Otto Hahn and Werner Heisenberg, who had worked on nuclear fission in Hitler's Germany and who were captured and taken to England in 1944 for questioning, when informed of the Hiroshima bomb during their confinement, could at first not believe it. In their estimation it was impossible for any nation to have separated enough of the low-abundance U-235 isotope for a bomb within a period of 3 to 4 years.

In the DIF method (Brief 17a), gaseous UF_6 is pumped through long pipes with porous walls possessing microscopic holes which slightly lighter $^{235}UF_6$ can slip through a little faster than slightly heavier $^{238}UF_6$. Thus the gas coming through the porous walls is a fraction more enriched in $^{235}UF_6$. By repeated recycling of the UF_6 gas through compressors in hundreds of stages, the U-235 fraction is gradually increased, the number of stages being determined by the desired final U-235 enrichment. In the electromagnetic or 'calutron' isotope separation method, uranium atoms are ionized and when passing through an electromagnetic field, heavier ions (U-238) follow a slightly different path than lighter (U-235) ions so they can be focussed at two different collection points. However this only works at very low pressures (a millionth of an atmosphere) and hence calutron throughputs are very low. They also require a lot more energy and cost per separated U-235 atom than in the DIF case.

Following the WW-II development of the gaseous diffusion process (DIF) for uranium enrichment, two other competitive methods have come to the fore, namely ultracentrifuge (UCF) enrichment and laser isotope separation (LIS). Brief 17 sketches these isotope separation concepts which use gaseous uranium hexafluoride. The latest LIS scheme is still in the development stage, but promises to reduce present uranium enrichment costs by a factor of five. The fully developed UCF method is currently favored over the DIF scheme since it consumes much less energy. Today Europe, Russia, and China all have operating UCF plants, while the US is replacing its old DIF plants from WW-II with a UCF enrichment facility. Commercial enrichment companies providing reactor-grade uranium for power plants are AREVA (France), URENCO (England, Germany, Netherlands), and USEC (USA). Enrichment services for power reactors are also available in Russia, China, India, and Japan.

In the UCF technique, gaseous UF_6 is swirled around at very high speeds, causing the heavier $^{238}UF_6$ to accumulate closer to the outside wall of the centrifuge and circulating downwards, while lighter $^{235}UF_6$ concentrates more inwards toward the center of the rotating cylinder and circulates upwards. By continuous removal of the upper and lower gas streams and by recycling through many centrifuge stages, enrichment of UF_6 is achieved.

In molecular laser isotope separation (MLIS), a supersonic free jet of UF_6 mixed with a carrier gas such as xenon, is coaxially illuminated by an infrared laser beam as it expands from a nozzle into a low-pressure chamber. Gaseous $^{235}UF_6$ in the jet is selectively excited by tuned laser photons, the selectivity being possible because of a shift between the absorption spectra of $^{235}UF_6$ and $^{238}UF_6$. Laser excitation of $^{235}UF_6$ suppresses it from forming dimers downstream as the jet cools down, and also promotes its migration out of the jet core. Non-excited $^{238}UF_6$ molecules on the other hand form UF_6:Xe dimers with xenon atoms as the jet cools itself by supersonic adiabatic expansion. Since the dimers carrying mostly U-238 stay closer to the central core of the jet flow, while $^{235}UF_6$ monomers fly out of the jet, the two

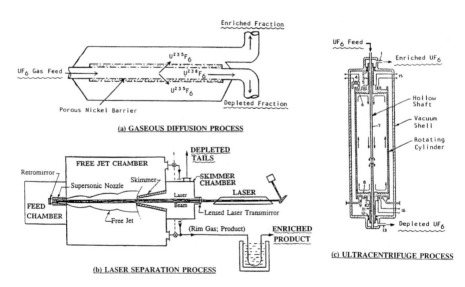

BRIEF 17. ISOTOPE ENRICHMENT SCHEMES FOR GASEOUS URANIUM HEXAFLUORIDE

streams can be separated by intercept of the free jet flow with a skimmer that sucks up the core of the jet but excludes the rim gases, which are pumped out separately.

The DIF technique requires hundreds of stages to enrich 0.7% U-235-enriched UF_6 to 3%, while a UCF plant can do this in tens of steps and MLIS in two to three stages. Brief 17 illustrates DIF, UCF, and MLIS unit separators. Besides a desirable high enrichment factor per stage which determines the number of stages, high throughput factors are also important. In the DIF enrichment process, operations are carried out at near-atmospheric pressures resulting in high throughputs, while in UCF and MLIS, gas pressures are less than a hundredth of an atmosphere. Thus throughputs per separator are much less for UCF and MLIS units compared to DIF, which means more separator units must be put in parallel for a given total throughput. Clearly, enrichment factors, operating pressures, energy consumption, and costs of capital equipment (and their maintenance) are among the main factors determining the economic preferability of an enrichment technique.

Besides the currently explored condensation-repression MLIS technique, another LIS scheme called AVLIS (Atomic Vapor LIS) was investigated in the 1970's, first by the AVCO Research Labs in Cambridge, Massachusetts, and later at the Lawrence Livermore Laboratory (LLL) in California. AVLIS laser-irradiates and processes hot (corrosive) uranium metal vapor instead of cold molecular UF_6 used in MLIS schemes. It requires an electron beam to evaporate metallic uranium in a high-temperature vacuum furnace. Three rapid sequential laser pulses selectively excite and ionize U-235 in the uranium vapor as it passes by electrodes, which deflect and collect the ionized U-235. While operation of a pilot unit in the 1980's succesfully demonstrated the technical feasibility of AVLIS, it has now been mothballed in favor of a more economic and more reliable UCF plant being readied by USEC for 2007 to replace its aging DIF plants. (USEC = United States Enrichment Corporation, successor of the former US AEC/DOE uranium enrichment enterprise).

5.2.2 Fuel (Re-)Processing

In a spent-fuel reprocessing plant, fission products are separated from the remaining uranium and from neutron-bred plutonium. Great precautions are taken in such plants to prevent accumulated plutonium from falling into the wrong hands. Purified separated plutonium, like highly enriched uranium, can be used for making nuclear weapons. The original plutonium present in a spent-fuel element has some isotopic components that make it unsuitable for use in 'burner' reactors, else newly formed plutonium could be directly burnt up in the latter. However fast or epithermal reactors can utilize this plutonium once it is separated from fission products.

When spent fuel elements enter a reprocessing plant, they are first cut open in a special room by remotely operated mechanical 'slaves' and manipulator arms. The contents are dropped in nitric acid solution which forms various chemical salt solutions with the uranium, plutonium, and lower-mass fission products (molybdenum, promethium, cerium, etc). The salts containing uranium and plutonium (and transuranics) are then removed first from the mixture by chemical exchange (using Tributyl Phosphate (TBP) solvent), fractionation, or other chemical extraction process. Finally valuable uranium, plutonium (and some transuranics) are retrieved for recycling into reactor fuels.

The remaining salt solutions containing fission products are treated to retrieve valuable radioisotopes used in nuclear medicine and other applications. The fission products mixture that is finally left has little commercial value and is heated and vitrified into solid glass-like marbles. These marbles are dropped into canisters which are shipped to a radiowaste disposal site such as Yucca mountain.

The annual amount of radiowaste produced by a typical 1000 MW(e) nuclear power reactor is about 250 kg (400 lbs). Including shielding and packaging, this amount of solid radiowaste can be hauled away in one railroad car or truck. In comparison, 300,000 kg (660,000 lbs) of ashes and cinders are produced by a coal-fired power plant of 1000 MW(e) output, which also releases 11 million tons (22 billion lbs) of globe-warming CO_2 gas into the atmosphere. In addition to CO_2, 55 million kg (120 million lbs) of SO_x gases and 27 million kg (59 million lbs) of NO_x gases, together with 240 kg (530 lbs) of mercury and 409 kg (900 lbs) of uranium are released annually into the air as entrained toxic materials.

In contrast with coal or natgas power plants, no gases are emitted into the atmosphere by nuclear reactors. They are absolutely non-polluting and should be applauded by environmentalists for this feature. Radiowastes from nuclear reactors are: (a) solid; (b) very modest in quantity; (c) valuable (in part) to nuclear medicine and biotechnology research; and (d) residual waste can be easily and safely contained for disposal. In contrast, the additional waste products from oil-, natgas-, and coal-burning power plants, aside from globe-warming CO_2 gas, are: (a) entrained gaseous dusts; (b) voluminous and large in quantity; (c) air-polluting, toxic, and useless; and (d) costly to separate from the main CO_2 exhaust.

5.2.3 Final Disposal of Radiowaste

As mentioned, useless fission-product waste (called 'fissium') which can be vitrified into solid marbles, is placed in steel canisters for final disposal. The canisters may typically be 20 cm (8 inches) in diameter and 2 to 3 meters (6 to 9 feet) long. Canister walls are made of 1 cm (3/8 inch) thick nickel-alloy steel. Only evanescent gamma and xray radiation (no particles) can escape from a canister. Originally each canister packs about 3 million curies of radioactivity which generates about 2 kilowatts of decay heat (equivalent to the power consumed by 20 lightbulbs). After shipment to a nuclear waste repository site the canisters are taken to underground vaults as shown in Brief 18. They are cooled there by circulation of filtered air drawn through chimneys and air ducts. Alternatively the canisters can be set into holes in the vault and be cooled by thermal conduction to the soil.

Previous studies of radiowaste disposal included canister storage in ten-meters (30-foot) deep water basins with water providing shielding and cooling by natural convection. Placement of radiowaste canisters on a basaltic deep-ocean floor was also considered (Brief 18). However more easily accessed land-based underground storage in abandoned mines or similar excavations are favored now. Use of gamma radiation from radiowaste canisters may become of interest for food preservation and sterilization of medical supplies (this is being researched). It is estimated that three abandoned mines, after conversion into vaults, can handle radiowastes of all US power plants for a thousand years under a breeder reactors regime.

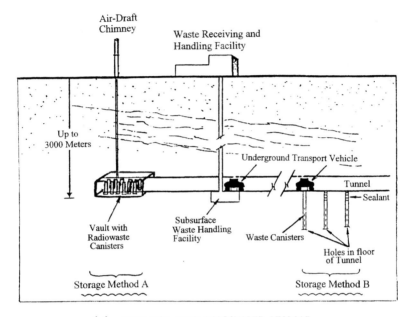

(a) SALT-BED MINE RADIOWASTE STORAGE

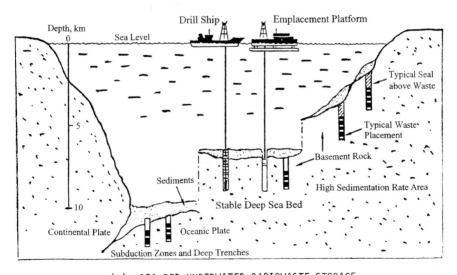

(b) SEA-BED UNDERWATER RADIOWASTE STORAGE

BRIEF 18. RADIOWASTE DISPOSAL SCHEMES

Suggestions to send radiowaste into space and into the sun have been made, but such proposed schemes are very costly and totally unnecessary. Fears that man-made radiowaste vaults constitute a long-term hazard for future generations are more emotional than real, if one considers that uranium and other natural radioactive ore bodies already exist on earth and have been there for millions of years. They are no less of a hazard to man than artificial ones. The amount of radiowaste produced is much less than many man-made hazardous materials that are routinely used. For example the annual amount of arsenic trioxide imported into the US for use in agriculture, is ten times larger than the annual reactor radiowaste that is produced by all US reactors. Arsenic trioxide is more toxic than decaying radiowaste and is stored above ground as opposed to underground radiowaste storage.

As time passes, radioactive waste products decay to stable non-radioactive elements, the shorter-lived species first and longer-lived ones last. Long-lived plutonium is a very valuable fuel and is removed as completly as practical from spent-fuel material in a processing plant. However a small amount (less than one half percent) may remain as a trace in the final fission product radiowaste and will get stored with it. Anti-nuclear critics state that because of the 24,000 year half-life of plutonium present in trace amounts in the radiowaste, a 500,000 year biological hazard is created. This statement sounds ominous but lacks any real substance, when one considers the fact thousands of tons of biologically hazardous materials like cyanides, nitrosamines, mercury, arsenic trioxide, etc., etc. have been and will be on earth for ever since their lifetimes are infinite. Yet we manage to handle and live with such materials even without having them locked away in deep underground vaults.

Critics also continue to bring up the fact that a liquid radiowaste storage tank hurriedly constructed during WW-II on the nuclear complex at Hanford (Washington state), leaked and had contaminated the surrounding soil. While succesful decontamination of this old WW-II storage tank required some effort, the actual degree of soil penetration was well below what was calculated in some hazards analysis reports. In the first few years of the nuclear age, the amount of radiowaste produced by nuclear facilities was small so one used the principle of infinite dilution instead of concentration to get rid of unwanted radioactive wastes. The dumping of small amounts of radioactive liquid solutions in the ocean (which is already naturally radioactive to a small extent), causes rapid dispersion and disappearance of any trace of enhanced radioactivity within tens of minutes. It is like adding a few grains of sand to the Sahara desert. The early practice was to dilute and dissolve weakly radioactive wastes into liquid solutions, and to store them in large tanks for later disposal at sea. With power reactors coming on line, the handling of more highly radioactive waste had to be switched from one extreme to another: concentration instead of dilution. Interim storage of radioactive liquid solutions still takes place at reprocessing plants to allow removal of useful medical radioisotopes and decay of short-lived species, before concentration and vitrification of long-life useless radioisotopes is undertaken.

Engineers see no difficulty with securely storing radiowastes underground in deep mines in salt-bed formations or similar excavations. Salt-beds are known to be geologically stable for millions of years. Fears have been expressed by critics that water flooding of such underground storage vaults (however unlikely) might cause nearby drinking water supplies to become contaminated and cause cancer deaths. Brief 19 shows possible long-term hazards associated with radiowaste storage as calculated by professor Bernard Cohen of the University of Pittsburgh under worst-case scenarios. He assumed stored radiowaste vaults could unintentionally be inundated with (ground) water. Assuming the canisters would rust through and radiowaste-contaminated water would seep into the ground, he estimated its possible arrival at a nearby well used for drinking water, as the radioactive species spread through the soil by diffusion.

Cohen's charts in Brief 19 show that fatalities due to drinking of possibly contaminated water are negligible, even if one assumes the conservative 'linear extrapolation theory' of induced cancers. (Actually the linear cancer induction theory has been proven incorrect; mild amounts of radiation are in fact beneficial – Ref 35). Of course it is unlikely that flooding and rusting of canisters would go unnoticed by future sentinels of a radiowaste repository. If signs of canister rusting should occur, future generations would probably re-jacket the canisters if they found them to pose a hazard. Maintenance of a disposal site is fairly trivial, and requires only monitoring and periodic patrolling. If water would somehow enter the vaults, it

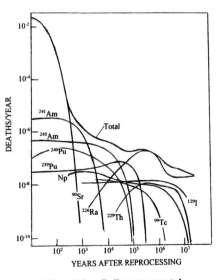

(a) Hazards from Radiowaste generated by U.S. Nuclear Power Reactors, assuming Salt-Bed Mine Storage, Water Flooding, Corrosion, and Seep-through.

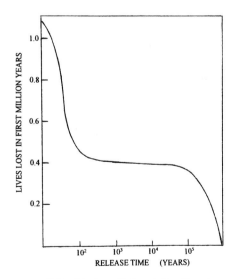

(b) Fatalities due to Radiowaste Seep-through in One Million Years as a Function of the Time required for Contaminated Water to reach Drinking Water.

BRIEF 19. LONG-TERM HAZARDS OF RADIOWASTE STORAGE IN SALT-BED EXCAVATIONS can be pumped out. But even if nothing was done, Cohen's calculations show that possible hazards to man are miniscule.

Concern has also been expressed about the possibility that terrorists might try to acquire radioactive material for use in a so-called 'dirty' but non-nuclear bomb. Any terrorist gang who would want to break into a radiowaste repository to steal canisters of radioactive waste for some evil purpose would have to bring a truck, winch, and special engagement equipment to retrieve any. Even if a gang was able to subdue the repository guards by guns or in a gunfight, they would mostly expose themselves to radiation and could do little harm to anyone else, should they succeed with such a heist. Damage from the explosion of a 'dirty' bomb comprises mostly mechanical blast effects. Radioactive materials are easily detected and a dirty bomb blast area is readily decontaminated with so-called 'rad-waste' solvents. Anyone not killed by the bomb's concussion but covered with radiodust can and should take a quick bath, shower, or swim to wash off radioactive particles. Any gamma radiation exposure from radioactive dust is evanescent and does not stick. If the terrorist's goals are to poison people, there are many poisonous chemicals available that would be more effective than radiowaste. In short, stealing radiowaste canisters is as pointless as recovering an old WW-II army tank from the bottom of the ocean for use in an armed robbery.

The end of the uranium fuel cycle has been solidly explored and engineered for several decades. After lenghty studies of possible sites (earthquake faults, water tables, etc), the US DOE (Department of Energy) selected, designed, and built a billion-dollar repository for the storage of high-level nuclear waste in the Nevada desert at Yucca mountain. The only problem with getting DOE's Yucca mountain repository into service has been the unreasonable opposition by anti-nuclear activists who keep court litigations going by interjecting trumped-up transportation safety concerns and other delaying tactics. Originally Yucca was to have opened in 2000. Because of the delays, nuclear power plant operators have resorted to temporary storage of fuel elements in swimming pools until Yucca will start to accept and process them. While this practice can be carried out safely from a radioactivity management point-of-view, and is checked out and approved by the NRC Nuclear Regulatory Commision), it would appear more prudent to have all reactor radiowaste stored at one well-guarded site instead of at a hundred reactor sites all around the country.

5.3. NUCLEAR FUSION

In contrast to nuclear fission which involves the heaviest known elements, in nuclear fusion a merging of two of the lightest elements occurs in which energy is released. Nuclear fusion was investigated long before fission by astrophysicists who tried to explain what caused the stars and our sun to emit so much energy for so long. From spectral and other observations it was known that hydrogen (H) and helium (He) had to be involved and that nuclear reactions had to produce the billion-year-long emissions from these nuclear furnaces. Some of the early physicists who studied nuclear fusion between 1915 and 1939 were A.S. Eddington, J. Perrin, W.D. Harkins, E.D. Wilson, R. d'E. Atkinson, F.G. Houtermans, C.F. von Weizsäcker, H. Bethe, G. Gamov, and E. Teller. These are authors quoted by S. Chandrasekhar, himself an astrophysics pioneer. Stellar fusion reactions take place at extremely high temperatures of millions of degrees where all elements exist as ions stripped from their outer electrons. This state of affairs is called a 'plasma', and plasma physics plays an integral role in fusion physics. Thus in describing nuclear fusion events, one uses symbols H = ^{1}H (proton), D = ^{2}H (deuteron), T = ^{3}H (triton) to respectively describe the hydrogen nucleus consisting of one proton of atomic mass 1 with unit positive charge, the deuterium nucleus comprising one proton and one neutron with total atomic mass 2 and unit positive charge, and the tritium nucleus with one proton and two neutrons having total atomic mass 3 and unit positive charge. The symbols ^{3}He and ^{4}He are used to indicate helium atoms stripped of their two electrons with respectively a helium nucleus containing two protons and one neutron with total atomic mass 3, and a helium nucleus with two protons and two neutrons yielding an atomic mass of 4. The ^{4}He ion is identical to an alpha particle identified in early radioactive decay research by madame Curie and others. With two neutrons and two protons, the alpha particle (i.e. a ^{4}He ion), is the most stable nuclear particle often emitted by decaying radioactive elements besides a neutron n or proton p ≡ ^{1}H.

To imitate solar fusion reactions on earth, it is necessary to confine a plasma with ionized species at temperatures of 10,000,000 degrees. This can and has been done by placing it in a strong magnetic field where positive and negative ions are constrained to travel helical paths around magnetic lines of force. By proper shaping of this field, charged nuclei can be confined for enough time to undergo fusion reactions. From the unraveling of fusion physics in the sun, and with improved data obtained from Cockroft and Walton accelerators in the 1930's, it was deduced in the 1940's that the following two nuclear fusion reactions are easiest to ignite at the lowest temperatures:

(5.2) $\quad ^2D + {}^3T \rightarrow {}^4He + {}^1n + 17.6$ MeV

and:

(5.3) $\quad ^6Li + {}^1n \rightarrow {}^4He + {}^3T + 4.8$ MeV

or overall:

(5.4) $\quad ^6Li + {}^2D \rightarrow 2{}^4He + 22.4$ MeV

Another possible fusion reaction with two different outcomes of almost equal probability is:

(5.5a) $\quad ^2D + {}^2D \rightarrow {}^3H$ (0.96 Mev) $+ {}^1H$ (2.88 MeV)

(5.5b) $\quad ^2D + {}^2D \rightarrow {}^3He$ (0.80 Mev) $+ {}^1n$ (2.5 MeV)

Released energies are in atomic units: 1 MeV = 1 Million electron-Volt = 1.6021×10^{-13} Joules. Reaction (5.5) requires higher ignition temperatures but may be active once fusion is initiated in a plasma.

Although intense research and development of controlled nuclear fusion has been conducted for the last 60 years, instabilities in electromagnetic confinement are formidable obstacles and exploratory costs are enormous. Alternative methods to ignite the above reactions using pulsed focussed beams of laser photons or particles that strike pellets of fusionable ingredients (called Inertial Confinement Fusion) are also being explored. The latest multinational collaborative effort is ITER (International Thermonuclear Experimental Reactor) which one hopes will demonstrate at least 'break-even' nuclear fusion. It is expected to be ready for experiments in 2008 but a site has not been chosen yet. Compared to nuclear fission, many believe nuclear fusion cannot compete economically for some time (at least 50 years), even if one can extract electric power from a device such as ITER. Since uranium fission can provide man with needed energy for 1000 to 2000 years, chances are good that technology needed for controlled nuclear fusion will have been conquered in that period of time.

Some mistakenly believe that nuclear fusion will not produce radioactive waste. Reaction (5.2) shows that fusion produces neutrons, and even though it is hoped that most will be captured by reaction (5.3), a good fraction will be absorbed by confinement materials of construction whose nuclei will thereby be transmuted and become radioactive. Also the size of a fusion power-producing reactor will of necessity be much larger than that of a fission reactor, so there is a large exposed area for neutron activation.

The main attraction of nuclear fusion is that there is an abundant supply of deuterium on earth. Even though deuterium's abundance is only D/H = 0.00015, the world's oceans hold approximately 100 trillion tons of deuterium, and a similar amount is estimated for ^{6}Li with natural abundance $^6Li/(^7Li + ^6Li) = 0.075$. Of course energy required for isotope separation of D and ^{6}Li must be less than the potential fuel energy in the isotopes. As in the case of U-235, because of the large amounts of energy released in nuclear processes, energies required for separation of isotopes are generally less than 0.1% of the harvested nuclear energy. With the extraction of deuterium from 1% of all the oceans, one calculates that 1 trillion tons of deuterium and lithium could provide the world with 10^{21} GJ of fusion energy. With a stabilized total world consumption of 10^{12} GJ/y, this supply could provide the world's energy needs for a billion (10^9) years, which happens to be about the remaining life of our sun.

While 'Hot Fusion' induced by high-energy nuclei was the main focus from 1945 till 1989, in 1989 M. Fleischman and S.J. Pons at the university of Utah claimed to have discovered so-called 'Cold Fusion'. In an electrochemical cell, where hydrogen is produced by electrolysis of water (Eq (4.1)), they believed to have observed the emission of nuclear fusion products. They used heavy water (D_2O) which generates D^+ ions (same species as in fusion) and employed a palladium electrode presumably generating reactions (5.5) on/in the palladium electrode surface. Because of the exponential 'Gamov factor' which gives the probability for two positive ions to penetrate the nuclear repulsion barrier between them, such a reaction is almost impossible at room temperature. It requires temperatures of 10,000,000 degrees kelvin for the D's to penetrate the mutual repulsive barrier and fuse.

A different fusion reaction between deuterium ions on a solid electrode (omitting + charge signs) might be:

(5.6) $\quad$ D + D : electrode $\rightarrow$ ^{4}He (23.8 MeV) + electrode

This reaction might explain why neutron emission rates reported by Pons and Fleischman were far below what would be expected on the basis of reaction (5.5b). Extremely energetic alpha-particles (^{4}He) possessing 23.8 MeV by (5.6), can generate neutron emissions in secondary nuclear interactions with intensities much less than what one expects if the primary reaction (5.5b) was involved. Unfortunately Pons and Fleischman (PF) did not measure possible ^{4}He or ^{3}He productions. The reaction:

(5.7) $\quad$ D + D + ThirdBody $\rightarrow$ ^{4}He + ThirdBody + 23.8 MeV

has been observed to occur in stellar plasmas at a rate one thousandth (10^{-3}) of that for reactions (5.5a) and (5.5b). The main reason for the low probability of (5.7) is that the laws of physics demand simultaneous conservation of momentum and energy in a two-particle encounter such as D + D. If only one outgoing particle is formed, this law would be violated. Only in the presence of a third body which can carry off balancing energy and momentum as in (5.7), is the reaction possible. In a gaseous stellar plasma, one out of a thousand two-body collision events takes place in the presence of a simultaneously colliding third body, which allows reaction (5.7) to proceed. If a D + D reaction took place on an electrode wall by (5.6) as in the PF experiments, the wall can act as the required third body provided it interacts resonantly with the D's and ^{4}He at the nuclear level. An explanation must still be found however to explain nuclear repulsive barrier penetrations at low temperatures on a palladium (or other) surface.

Many who tried to duplicate the 1989 Pons-Fleischman experiments failed to get the same results, except for a few who claimed to have also seen some nuclear emissions (neutrons, ^{3}He, and ^{4}He) and release of excess heat. Most of these emissions occurred in occasional spurts during extended (week-long) D_2O electrolyzing runs. Numerous theories have been proffered to explain a possible nuclear process, while just as many have been proposed to give a non-nuclear chemical explanation of these observations. Presently the majority of nuclear physicists are skeptical and believe that spurious observations of 'cold fusion' are due to nuclear background interactions (cosmic rays?) or because of inaccurate and/or inadequate measuring instruments. To date (2004), no absolute scientific proof acceptable to all physicists has been provided yet for PF-like cold fusion phenomena. Nevertheless cold fusion research is still being pursued by believers.

CHAPTER 6

SAFETY CONSIDERATIONS IN NUCLEAR OPERATIONS

While some aspects of reactor safety procedures were already discussed in Subchapter 5.1, here we give a more general review of safety measures that have been instituted for reactor and nuclear fuel handling operations. We start with discussing the nature of nuclear radiation and its effects on man and biomatter. Next we consider radiation dose measurements and tolerable exposure levels, followed by a review of radiation protection measures and safety assurance in reactor operations. Finally, some typical mishaps in reactor operations and fuel handling are discussed and a review is given of serious nuclear accidents that have occurred since the beginning (1945) of the nuclear age, including the Three-Mile-Island (TMI) and Chernobyl reactor meltdowns in 1979 and 1986.

6.1. NATURE OF NUCLEAR RADIATION

The major nuclear emissions in an operating reactor consist of alphas, betas, gammas, and neutrons. Alphas are highly energetic helium ions ($^4\text{He}^+$), while betas are nothing but very high-energy electrons. Like fission fragments, most alphas and betas slow down and are completely stopped after a few millimeters (0.04 inches) of travel in solids, and after 2 to 6 meters (6 to 18 feet) in most gases at atmospheric pressure. As they slow down their kinetic energy is converted to heat in a fuel element. It is impossible for fission fragments, betas, or alphas to fly out of the reactor core, through the neutron reflector, and out through the gamma shield.

Gammas are the only radiation species that can pass through the core and reflector of a reactor. For this reason a special gamma shield is placed around the reactor core and reflector to stop them. Gammas are massless evanescent high-energy photons like xrays and light. The typical 1 MeV gammas emitted in fissions can travel through 150 cm (60 inches) of water or 10 cm (4 inches) of lead before they are attenuated 100-fold. Although they have no mass, gammas do have energies which generally lie between 0.1 and 10 MeV (see Sec 5.3). This compares with visible light whose photons have energies between 1.9 and 3.6 eV approximately, that is one millionth less than gammas. All photons, gammas, xrays, and light vanish once they have been absorbed by matter or flown away into space. Their identity ceases and they do not stick (they are evanescent). The heavier the shield material is, the smaller the gamma's stopping length is. Typical lead shields are 20 cm (8 inches) thick, while (less costly) heavy concrete reactor shields are 1 to 2 meter (3 to 6 ft) thick. The exact shield thickness is determined by the criterion that the radiation dose level at the outside of the reactor shield must be 2.5 mr/h or less (see Subchapter 6.3).

Neutrons have approximately the same mass as positively charged protons or hydrogen nuclei (the lightest of all elements) except they have no charge and are neutral. Neutrons can therefore fly through a solid material like a gas. In uranium fission, neutrons have at first very high velocities when they are born, but as they diffuse through the reactor core they collide and slow down. The lighter the atoms the neutrons collide with, the more quickly they are thermalized. Since thermal neutrons promote fission better than fast ones, a neutron thermalizing material or 'moderator' such as water is dispersed through the core which otherwise contains primarily fissionable uranium or plutonium. The advantage of water as moderator is that it can also function as the reactor coolant. A neutron reflector placed around the core insures that most neutrons escaping from the core are reflected back into the core. To stop any remaining neutrons that might pass through the reflector, some extra boron (or other high neutron-absorbing compound) is often mixed in with or added to the gamma shield to bring neutron levels down to acceptable safe levels outside the gamma shield.

Nuclear radiation emitted by fissioning uranium and radioactive fission products in a reactor cannot travel very far, and their intensity drops off with the square of the distance from the reactor, aside from being considerably attenuated by the reflector and shielding. However radiation from so-called 'fall-out' often mentioned in connection with nuclear explosions, has a much farther reach. Nuclear fall-out consists of airborne particles or dust of pulverized radioactive fission products released as a cloud into the air from a nuclear bomb explosion or a Chernobyl-like fire from a reactor without a containment vessel (Subchapter 6.6). Such a cloud is carried downstream by the wind and can 'rain' on a populated area causing widespread radioactive contamination. In this case the radiation sources are in the radioactive dust particles taken far from their originally contained place of birth.

The fall-out of radioactive particulate matter, if not inhaled or ingested, can be rendered harmless by people covered with it by washing it off. A quick shower or swim is recommended in this case; of course it is even better to get out of the path of a fall-out cloud if one is warned by radio where it is going. It is extremely unlikely that any individual would receive an acute lethal gamma burn from fallout that covered him. Fall-out particles are clusters of mostly metal-oxides of radioactive fission products of uranium mixed with oxides of structural fuel element materials such as zirconium, steel, aluminum, and silicon. Since alpha and beta particles are stopped by a millimeter or so in most solids, dust particles provide some self-shielding, while betas that reach a person's skin are stopped in the skin and cannot reach internal organs. Only gammas emitted by the fall-out can provide a whole-body radiation burn with some body penetration, if a person is totally covered with fall-out.

If fall-out is inhaled and absorbed by the lungs or enters the digestive tract by eating or drinking it, a more serious situation can occur. There are three main 'bad actors' among the various radioactive daughter products of uranium fission which can be selectively absorbed by and lodged in human organs. As listed in Brief 20, these are iodine-131 and -129, strontium-90, and cesiun-137. Fortunately iodine-131 which collects in the thyroid, decays relatively fast. On the other hand, Iodine-129 (like the K-40 in our blood) lasts "forever", but delivers a very low dose because of its long life. Sr-90 and Cs-137 can last a long time unless metabolized or driven out before lodging in body parts. Short-lived radioactive species present in fall-out decay away to stable elements, and if inhaled or ingested are usually eliminated by the body.

BRIEF 20. 'BAD' RADIOSOTOPES FOR HUMANS AND MAMMALS

I-129 with a half-life of 17 million years → Thyroid (very low level)
I-131 with a half-life of 8 days → Thyroid
Sr-90 with a half-life of 28 years → Bones
Cs-137 with a half-life of 30 years → Cells

In summary, gamma radiation emanating from an operating reactor (if any) is evanescent. It does not 'stick' as some mistakenly believe. Only fall-out dust from a nuclear explosion can 'stick' to a person's skin. It can be washed off if not inhaled or ingested. If someone inadvertently walks through fall-out rain, he/she should breath through a wetted handkerchief or gas-mask if available, to prevent dust inhalation.

6.2. BIOLOGICAL EFFECTS OF NUCLEAR RADIATION

Many people are unaware that nuclear radiation is also emitted by nature and is all around us. It comes from the heavens (cosmic radiation) and from naturally radioactive minerals on the earth surface that emit it. Life on earth has evolved in this radiation climate for eons, and all bio-organisms have learned to tolerate or make use of it. When a gamma, xray, or ultraviolet photon strikes a bio-organic molecule in skin or in an internal organ, it will knock off an electron, that is it 'ionizes' the biomolecule. Biomolecules like DNA (Desoxyribo Nucleic Acid) and RNA (Ribo Nucleic Acid) are very large and contain thousands of atoms and electrons per molecule. Brief 21 illustrates the electron clouds around an atom and a simple molecule. The response to removal of an electron from a large biomolecule is that the electron cloud surrounding it will quickly refill the gap, while the loose electron after some migration, reattaches itself to another large biomolecule with thousands of electrons. Because of the immense 'sea' of electrons around biomolecules, one displaced electron usually produces hardly any effect.

Because man's body contains a lot of water, most other entities besides biomolecules that a gamma encounters in traversing through tissue are water (H_2O) molecules. Ionization of water usually results in the formation of a hydroxyl radical (OH) or hydrogen peroxide (H_2O_2). These molecules can attack a biomolecule and 'denature' (= kill) it. However many water ionization events result in a return of the liberated electron to the ionized water molecule without causing secondary destructions.

The above processes take place continuously on our skin when we expose ourselves to sunlight, which consists of ultraviolet (UV) and visible photons. The only difference between a UV and a gamma photon is that a UV photon can only remove one electron fom a molecule, while a gamma photon can knock off many electrons along its path, the distance between knock-off points being usually quite large (millimeters to centimeters). Thus a gamma photon can produce 10 to 1000 distributed ionizations in traversing human tissue. One might conclude from this that a gamma burn is more dangerous than a solar UV burn, but in reality the number of UV photons from the sun that bombard our skin is a billion times larger than the number of gammas that hit and pass through us from a typical gamma source. Clearly, one has to take into account both the number of photons per second (i.e. the flux), as well as the ionization strength of each photon. Tolerable radiation exposures are discussed in Subchapter 6.3 together with units commonly used for measuring radiation dosage.

Like gammas, strikes by high-energy betas or alpha particles also cause ionization, and thus electron stripping. However compared to gamma-produced ionizations, the ionization tracks of alphas and betas leave a much higher concentration of effected biomolecules, that is the electrons knocked off by alphas and betas are much more closely spaced. This also means that alphas and betas propagate only a millimeter or so, and are completely stopped by the human skin if one deals with an outside source. The effect is not much different from what a person would receive by a UV sunburn.

Internal organ exposure to alphas and betas is possible only if a radioactive particle that emits alphas or betas is inhaled, swallowed, or injected. In this case one finds that damage of biomolecules by betas and alphas can be more severe since ionizations are closely spaced. For heavy alphas it is necessary to assume a relative biological effectiveness (RBE) factor of RBE $\approx$ 10 to account for the higher ionization densities and its effects. That is, one alpha can do as much damage inside tissue as ten betas. For betas RBE $\approx$ 1 is usually assumed although some radiologists estimate that RBE $\approx$ 1.5 for betas with energies above 1 MeV is probably justified depending on what is irradiated.

In therapeutic nuclear medicine, advantage is taken of the high-density deposition of ionizing energy by betas and alphas. In this case, beta- or alpha-emitting radioisotopes are synthetically incorporated in certain pharmaceuticals which when taken internally, are adsorbed by tumors. Of course in nuclear medicine one

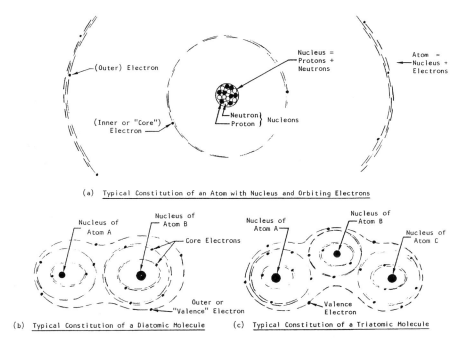

(a) Typical Constitution of an Atom with Nucleus and Orbiting Electrons

(b) Typical Constitution of a Diatomic Molecule

(c) Typical Constitution of a Triatomic Molecule

BRIEF 21. MAKE-UP OF A NUCLEUS, ATOM, AND MOLECULE

intentionally wants to overdose a selected target (the tumor) while one tries to minimize possible radiation damage to non-malignant organs. Using sufficiently strong doses, betas or alphas can destroy tumor cells. Nuclear treatment is ideal since betas and alphas can travel only a a fraction of a millimeter in tissue as they slow down, thus avoiding damage outside the selected tumor. The main problem is to get a therapeutic radioisotope or radiopharmaceutical only in the tumor cells and not elsewhere in the body. For this reason radioactive 'seeds' or microspheres are sometimes implanted in the tumor, as in the case of prostate cancer.

While most electron knock-offs from DNA molecules by gammas are assimilated by irradiated tissue as described above, now and then a collision occurs that causes a DNA or other biomolecule to break up or causes the molecule to restructure itself. These events are rare on an individual basis but do happen regularly if billions of gammas, betas, or alphas are continuously traversing biological tissue. Occasionally they can lead to a rearrangement in a biomolecule which might be the beginning of a possible cancer-forming mutation or carcinogen.

Since the radiation interaction process is a random one, the probability of producing a 'bad' atomic rearrangement from lesions in a biomolecule is proportional to dose. It therefore has become practice to assume 'linear scaling' of lesions or cancer formation rates with radiation dose, although this has not been proven. Biochemistry professor T.D. Lucky at the University of Missouri in Columbia has collected statistical data that indicate low-level radiation is actually beneficial to man [called 'hormesis' – Ref 35]. We shall further argue below that cancers due to 'bad' lesions might be proportional to the square of radiation dose and not linearly. We also show that it is more likely that mutations in DNA are induced by localized atomic-scale electric fields of non-regular invasive groups of molecules or atoms rather than by radiation. This refutes the assumption by some anti-nuclear activists that each cancer has its origin in a radiation interaction[9].

Before discussing possible cancer formation pathways initiated by radiation effects, we need to point out that the wound from a knife cut, a bruise, or a burn from a fire or hot stove on a person's body also produces a large number of lesions in quadrillions of molecules. The act of driving the sharpened edge of a sharp metal through flesh causes the electrons on the metal edge to break biomolecular bonds by the trillions. The end effect does not differ from the atomic displacements caused by alphas, betas, or gammas, except that the concentration of actual lesions is higher in the case of a cut, bruise, or burn. The body cannot tell the difference whether these lesions were caused by betas or gammas, or by knife-edge electrons. After defining radiation dose in Section 6.3, we shall show in fact how one can relate micro-scratches or knife-cuts to radiation exposures levels.

Probably the most frequent biomolecular mutations and lesions in the body are caused by unnatural impurity atoms or molecules taken in via food or drinks, and/or created by a bacterium or virus. When lodged in some internal organ or other part of the body, the intermolecular force fields of such 'foreign agents' can cause bond angle rotations, bond breakages, helicity changes, or a 'cross-over' of components during a DNA replication process. Particularly atoms or molecules with strong dipole moments may influence and possibly distort a biomolecule that is contacted. That chemical agents can cause biomolecular rearrangements resulting in cancer forming mutations is well proven. Many chemical 'carcinogens' have been identified and are now banned from food or drinking water. To date only leukemia has been proven to be induceable by radiation. In summary, although radiation can be a cause, the majority of cancers appear to be of biochemical origin, *not* radiation-related.

The chain of events that may lead from a single lesion or re-arrangement in a biomolecule to a cancer or genetic defect is believed to be initiated by a faulty replication of a 'healthy' biomolecule in an organism. Since biomolecules are repeatedly replicated, it takes but one wrong templet to build up a generation of defective or cancerous biomolecules and cells. Fortunately conditions that allow replication are rather restrictive and not every rearranged or broken biomolecule is

---

[9] The 'First Law' of antinuclear activist Gofman claims that all cancers are caused by radiation (Chap 2 of 'Population Control through Nuclear Pollution', by A.R Tamplin and J.W.Gofman; Nelson-Hall Co, Chicago). If this 'law' (called Gofman's 'First Flaw' by some) were correct, one deduces that every person on our planet should have died already ten times over from cancers due to cosmic radiation

replicated (most are eliminated). Nevertheless there are some replications that can cause a problem occasionally.

When a biomolecule is broken up in an ionizing radiation event, it is unlikely that the broken-off portion can replicate all by itself. It might re-attach itself to its broken-off partner in which case there is no net change (except for possible bond-angle changes). Or it attaches itself to another broken-off biomolecular part to form a modified molecule that may or may not be replicable. Replicable recombinations of two broken-off molecular pieces might lead to multiplication of faulty genetic material and thence to cancer. If cancer-forming mutations are formed in this manner, cancer rates would be proportional to the square of the radiation dose, not linearly. If proportional to the square of radiation dose, the effect of low-level radiations would be vanishingly small.

If instead of a recombinaton requirement, it is sufficient to induce a bond-angle or helix-angle change in a biomolecule to create a mutation, the cancer versus radiation dose would be linear since such changes can be effected by one ionizing interaction. At present there is insufficient information available to determine whether the relation is linear or quadratic, and it is conceivable that both postulated mechanisms are operative. The most important result from our review is however the fact that the majority of cancers are induced by chemical or biochemical (virus or bacterium) agents, not by radiation. As mentioned, recent research has indicated that low levels of radiation (enjoyed by people living at high altitudes) promotes health rather than worsens it. Like heat, a little bit from a heating pad can be healing, but the heat in a blast furnace would incinerate a person.

So far we only considered alphas, betas, and gammas. Neutrons produced in reactors can also induce mutations but in an entirely different way, since they can transmute atoms into other elements upon absorption. However they cannot travel far from the reactor and are metastable, decaying in twelve minutes. Neutrons are only plentiful inside a reactor. Unless neutron flux levels are extremely high, the biological effect of neutrons is similar to that of gammas but with $RBE = 3$ to 10.

In summary, nuclear gammas, betas, and alphas, like ultraviolet radiation from the sun, can cause: (a) ionizations with no final biological effect; (b) ionizations leading to denaturation of biomolecules repairable by the body; (c) unrepairable lesions of biomolecular bonds (less than one out of a million interaction events). In the latter case there is a small chance that a replicable biomolecular mutation is created which can lead to cancer, but most often the odd molecule is rejected and eliminated by the body. Whether the probability of cancer inducement is directly or quadratically proportional (or in-between) to dosage is still uncertain. If quadratic, the effects at lower dose rates should be much less than what is assumed on the basis of a linear extrapolation of high-level dose effects (where they can be measured). In conclusion, comparing possible cancer inductions by natural or man-made radiation to cancer inductions by chemical agents, the risk of getting cancer from chemicals in foods, drinks, and air, appears many times more likely than that from brief exposures to natural or man-made radiation.

6.3. RADIATION DOSE MEASUREMENTS AND TOLERABLE EXPOSURES

To determine what radiation levels are safe, one must first define a unit of radiation dosage. This unit is the 'rad' which is defined as 100 ergs = 10^{-5} joules of ionization energy transferred per gram of effected body tissue by radiation particles. In formula form:

(6.1)   1 rad = 100 ergs (= 10^{-5} J) of deposited ionization energy per gram of effected tissue

This definition still does not mean much without a comparison or reference as to how much or how little damage a rad does to man and how one measures it. This is discussed in what follows.

The amount of radiation energy deposition, the rad, must be multiplied by the relative biological effectiveness factor or RBE, before one can speak of a biologically meaningful dose. Different energetic particles have in general different RBE's. The RBE for gammas and betas is RBE = 1, for alphas RBE = 10, and for neutrons one assumes RBE = 3 to 10. When the rad is multiplied by the RBE, one obtains dosage in terms of rem (roentgen equivalent man):

(6.2)   RBE · rad = rem = 1000 mr (millirems)

Thus if a source of alphas deposits 100 ergs per gram in tissue, it injects a physical dose of 1 rad and a biological dose of 10 rem, since alphas have an RBE = 10. Since the RBE for betas and gammas is RBE = 1, for all practical purposes the rem and rad are equal for betas and gammas.

To get a better feel for the rad or rem, it is useful to compare rad damage with body damage created by a knife cut. If we define a 'micro-cut' as a 1 millimeter deep by 2 centimeter long knife cut or scratch on the skin of one's hand, one can show that the sharp knife edge causes approximately 3×10^{14} severances of biomolecular bonds in the knife wound. If one assumes that on average each bond breakage required 10 eV, the total energy dissipated in the wound would be 3×10^{15} eV = 4800 ergs. If it is further assumed that a hand weighs approximately 480 grams, the damage to the hand spread over the whole hand would be 4800/480 = 10 erg/gm or 0.1 rem = 100 millirems:

(6.3)   1 micro-cut = 1 mm deep by 2 cm long knife-cut ∼ 100 mr (on one hand)

It is clear from this analysis that it matters greatly whether tissue damage by radiation or knife-cuts is for the whole body or partial body. If we had allocated the knife-cut damage to one finger with a weight of 48 grams, the equiavelent rad damage would have been 1000 mr. Thus if one holds one's hand over a radiation source (e.g. Co-60) but the rest of the body is shielded, one should reduce the locally measured dose on the hand by the factor [mass of hand]/[mass of whole body] to arrive at the whole-body dose. Tissue damage from a directed beam of radiation which illuminates an area less than that of a person's body should also be adjusted accordingly. In measuring and recording radiation exposures, it is usually tacitly assumed that the observed radiation was permeating all of the space around a person. It is important therefore to record if a radiation exposure is 'whole-body' or 'part-body'. If not indicated, it is assumed to be whole-body.

It has been found from experiments with rats and many tests that a man working near a source of radiation can be exposed to about 1 rem = 1000 mr per week without noticeable effects. Assuming a 40-hour work-week this means an exposure rate of 1000/40 = 25 millirem (mr) per hour. For conservative reasons, the International Commission on Radiological Protection has added another factor of 10 and recommends that the so-called maximum permissable exposure (MPE) be set at:

(6.4)   MPE = 2.5 millirem/hour = 2.5 mr/h = 5 rem/y (2000 hr work-year)

Nuclear reactors are therefore designed with radiation shields such that the highest radiation level is not more than 2.5 mr/h in areas where men may walk near a reactor. Actually reactor operating personnel rarely stay for extended periods of time at such locations and their usual exposure is much less than this. Nevertheless the law requires that exposure levels do not exceed 2.5 mr/h in areas where personnel can move about in a reactor facility.

Directly ouside a nuclear reactor building, radiation levels are much less than 2.5 mr/h because of the inverse-square-with-distance attenuation. That is if 10 meters (30 feet) from the reactor core the level equals say 2 mr/h, one finds that at 100 meters (300 feet) the level is $(10/100)^2 = 0.01$ times less or 0.02 mr/h. If one is 1000 meter (3000 ft) from the reactor core outside the gates of the nuclear reactor plant, the level has dropped to 0.0002 mr/h which compares with 0.01 mr/h of background radiation one receives from cosmic radiation. If there are intervening structures (and there almost always are), the radiation levels outside the plant are even lower and undetectable.

To improve our understanding of what these radiation levels mean, it is instructive to compare them with naturally-occurring forms of radiation. As mentioned, the cosmos bombards us continuously with approximately 102 mr/y, while the average person receives about 72 mr/y from medical xrays and 59 mr/y from naturally radioactive potassium (^{40}K) and carbon (^{14}C) which are always present in our bodies (everyone is slightly radioactive!). Someone living on granite rock (Manhattan Island), or near a stockpile of coal receives another 1 to 10 mr/y of radiation from radon, a radium decay-chain element. People living at high altitudes like Denver, Colorado, receive an additional 50 to 100 mr/y. Finally the sun can give each person a skin dose of an astounding 1,000,000 mr/y from solar UV photons if he stood in the sun all year long (staying out of the direct sun for long times is highly recommended!). Brief 22 summarizes the radiation exposure each person gets from natural sources.

Clearly man has been continuously exposed to different natural radiation sources for millions of years, and succesfully evolved in this environment. Recent studies show that small amounts of radiation are actually beneficial (called 'hormesis'). Although other factors may be involved, people living at high altitudes like in Colorado or the Andes, are on average healthier than plain dwellers.

BRIEF 22. NATURAL RADIATION EXPOSURES

A. <u>CONTINUOUS WHOLE-BODY RADIATION</u>
Cosmic Rays	—	102 mr/y
Medical Xrays	—	72 mr/y
Potassium-40	—	55 mr/y
Carbon-14	—	3.6 mr/y
Global Fallout from 1950–1960 Atmospheric Nuclear Tests	—	3 mr/y

B. <u>MISCELLANEOUS SPECIAL SOURCES</u>
One Airplane Flight	—	1 to 10 mr
Coal Pile Radiation	—	50 to 100 mr/y
Granite Rock Radiation	—	30 to 80 mr/y
Living in Denver, Colorado	—	50–100 mr/y

C. <u>SKIN IRRADIATION</u>
Solar Ultraviolet (UV) — 1,000,000 mr/y $\approx$
$\approx$ 114 mr/h
for average person

Aside from the safe maximum permissable exposure of 2.5 mr/h, on the other extreme one has found that a dose of 500,000 mr = 500 rem is generally lethal to a man if received all at once:

(6.5)   Accute Lethal Dose = 500,000 mr = 500 rem (instant whole-body)

If one receives this total-body dose of 500,000 mr in a few seconds, there is a 50-50 chance that death will follow. This high level corresponds to throwing a lobster in boiling water or a person into a furnace. The heat generated by this heavy dose kills a super-critical fraction of living cells in the body which the body is unable to replace fast enough.

Unless death is instant, victims who got a dose of 500 rem can sometimes still be saved by administering injections of transplant bone marrow. An excessive dose of radiation destroys most of a person's bone marrow which is needed to manufacture new bloodcells. Normally the body would try to repair itself after a trauma, but if the bone marrow 'factory' itself is put out of commission, new blood cells cannot be made any more at a sufficient rate to keep up with the demand. By providing bone marrow artificially to a victim from a bone marrow bank (like a blood bank), his body can build up a new bloodcell supply until the bone-marrow factory can rebuild itself. Two men from the 1958 Yugoslavia reactor accident, were saved in this manner after they were flown to Paris and given bone-marrow injections for a few months till their bodies resumed bloodcell manufacture again. However application of bone marrow transplants to Chernobyl victims in 1986 show that this treatment is only effective for a narrow range of exposures between 100 and 400 rem (see Section 6.6).

Some remedies against inadvertent internal radiation exposure have been developed in the form of radioactivity elimination tablets. In cases where a victim accidentally inhales or swallows fission-generated radioactive iodine-131 which tends to lodge in the thyroid, one has found that taking pills containing potassium iodide (KI) are a preventive. If taken quickly after the accident, KI will saturate the thyroid with its non-radioactive iodine and block additional radioactive iodine-131 (half-life = 8 days) from entering. The radioactive iodine-131 is thus prevented from accumulating in the thyroid and is eliminated by the body. Iodine-131 is the longest-lived of several radioactive bio-sensitive iodine isotopes produced in nuclear fission. Accidental inhalation of iodine-131 could happen for example in handling a gas-leaking cracked fuel rod, or if one drank contaminated milk from cows that grazed on grass covered with radioactive fall-out. After two months (8 half-lives), I-131 is essentially decayed away. Administration of potassium iodide pills to 130,000 residents within a radius of 30 km exposed to Chernobyl radio-iodine fall-out (Sec 6.6), proved quite successful with no side effects. Pills scavenging radioactive Sr-90 and Cs-137 in the body to help eliminate them are reported to be still under evaluation.

Gamma and beta radiation doses can be measured by a number of instruments, the most common one being the radiation survey meter which uses a gas-filled Geiger-Müller (GM) tube. Any ionizing particle passing through the thin metal window of the tube into the gas generates free electrons whose current is measured. The survey meter is calibrated against a source of known strength and indicates the radiation level on a dial in proportion to the ionized electron current. Some meters also convert the current strength to sound, making a crackling noise proportional to radiation intensity.

Workers in nuclear facilities carry 'dosimeters' or 'film badges' clipped to their clothing. The dosimeter or 'radiation pencil' comprises an electroscope device that measures integrated radiation exposure and is checked daily, while the film badge exposes a special nuclear-radiation-sensitive film which is checked for exposure effects at least once a week (or sooner if there is cause for a possible radiation overexposure). The resident health physicist is usually responsible for regular monitoring, log-keeping, and maintenance of personnel dosimeters and film badges.

Instead of the historical rad and rem, one has now also defined the 'Gray' and 'Sievert' as:

(6.6) $1\,\text{Gy}\,(\text{Gray}) = 100\,\text{rads} = 1$ Joule energy deposited per kilogram irradiated tissue

(6.7) $1\,\text{Sv}\,(\text{Sievert}) = \text{RBE} \cdot \text{Gy}\,(\text{Gray}) = 100\,\text{rem}$

These units follow the international mks system (mks = meter-kilogram-second).

In therapeutic nuclear medicine where tumors are irradiated, quoted doses are *not* whole-body as is customary in reactor operations, but instead apply only to the kilograms of tissue in the tumor. Thus doses of 5 Gy (500 rads) which would be lethal if it was a whole-body exposure, are not uncommon for killing a cancerous tumor. However the radiation energy is focussed on and/or restricted to the tumor only. The milli-Gray (mGy) = 0.1 rad or the milli-Sievert (mSv) are often used as basic units in diagnostic applications where low doses of a nuclear tracer are injected to follow a biological process.

6.4. RADIATION PROTECTION IN REACTOR OPERATIONS

Neutron reflectors and gamma shielding in reactors are designed to ensure minimum radiation exposures to reactor operating personnel. Compliance with safe reactor design rules and regulations are checked by the NRC (Nuclear Regulatory Commission) when an application is filed for reactor construction, which is repeated after a license is issued and the reactor is built. After reactor start-up, there is still another checkout. Upon commencement of routine nuclear power generation, the NRC appoints a health physicist (HP) to be in residence at the reactor, who monitors and records radiation levels around the reactor and who verifies there is continued compliance with all safety features specified in the NRC-issued reactor operating license. The HP makes sure that personnel radiation levels in the reactor building as well as throughout the reactor site are always below maximum allowed levels (see Subchapter 6.3), and that all radiation monitors and alarms are in working condition.

The resident HP usually reports directly to the NRC and can bypass the reactor plant operations manager if he chooses. This ensures that reactor safety is never compromised in favor of economic or other considerations. The job of a reactor health physicist is comparable to that of an air traffic controller at an airport. Careful selection of a mature responsible health physicists is an important task of the NRC. Although independent, an HP must not misuse his powers and must promote safe operations through teamwork and good relations with reactor management.

The most often used reactor shielding materials are lead, concrete, and water. Besides a neutron reflector of water, most reactors have a shield of concrete and/or lead around it to stop gammas. Betas, alphas, and fission fragments are completely stopped in the solid reactor core, and only neutrons and gammas an escape from it. Most neutrons in a power reactor are reflected back into the core by 30 cm to 100 cm (1 to 3 ft) of water since they are valuable in maintaining the fission chain reaction and as few as possible are allowed to escape. The thickness of the gamma shield is usually about 30 cm (1 ft) or more of lead, and/or 1 meters (3 ft) or more of concrete.

Around the basic reactor is the 'containment vessel', which surrounds the 'reactor pressure vessel', the reflector, and shielding. The containment vessel is usually a enormous sphere or cylinder of 5 cm (2 inch) thick steel whose sole purpose is to keep all radioactive vapors and debris inside it in case of an accidental reactor meltdown. There is also an emergency core cooling system (ECCS) which provides backup coolant for the core, should the regular cooling system fail. Because of 'after-heat' from radioactive decay products, core cooling needs to be continued even after a reactor scram, and this is done by the ECCS if the normal system fails. In case of a reactor scram, spring-driven control rods which are designed to be fail-safe (Section 5.1.1), drop immediately back into the core causing cessation of any further uranium fissioning.

In the reactor building, a number of radiation monitors are mounted on walls or on the ceiling which will sound an alarm if they detect radiation above a certain threshold value. In ventilation ducts and air circulation passages, additional radiation monitors are usually situated at strategic locations. For the sake of safety, several redundant monitors are present which are checked weekly to insure their operability. The power supply for each monitor is automatically switched to a standby battery in case of an electric power grid failure. These batteries are kept charged and checked continuously. Besides gas-filled ionization chambers, solid-state semiconductor detectors have been developed and are used today to perform the basic sensing function in radiation monitors.

6.5. MISHAPS AND MALFUNCTIONS IN REACTOR OPERATIONS

Nuclear reactor power plant operations, like all other human undertakings, occasionally experience a malfunction or other disturbance that leads to an accident. Many people ask if reactor accidents are tolerable and reasonable when compared to other human endeavours. We shall show that the risk of running a nuclear plant is no different and probably less than that of an oil-, natgas-, or coal-fired power plant or chemical industry. Can reactors be operated safely? The answer is a definite yes.

In the fifty-year history of nuclear reactor operations, one finds that most accidents in nuclear power plants or nuclear-powered submarines are of a non-nuclear nature and do not breach nuclear safety features. The most common accidents which cause a reactor to be shut down are malfunctions in electronic circuitry, electrical systems, and human error. Because reactor operating procedures require that all main reactor electronics be fully functional all the time, a temporary reactor shutdown is sometimes necessary for electronics repair. Such shutdowns cause a costly temporary loss of electric power generation, so in modern reactors, two or more parallel or redundant electronic systems are provided which can take over from each other should one system fall out. This allows quick modular repair of a failing component without disruption of power plant operations.

In addition to simple electronic component failures, other typical non-nuclear accidents have been electrical fires, the failure of a valve in a coolant system, or the cracking of a pipe. Typical nuclear accidents that occurred in earlier phases of the nuclear power program were the swelling and cracking of cladding or encapsulation materials in fuel elements which resulted in the release of some radioactive fission products into the coolant. None of these accidents were catastrophic or caused major radiation overexposure to personnel and most were satisfactorily repaired. Many early mishaps led to significant improvements of component design to avoid similar occurrences in the future.

Radioactivity build-up in reactor coolants is quickly detected by monitors that sound an alarm and, if severe enough, initiate a reactor scram. The contaminated coolant is usually filtered so radioactive products can be removed. Still in a 1957 reactor accident in England, some semi-volatile iodine-131 fission product escaped into the atmosphere due to a sudden crack in an air-cooled fuel element. It temporarily contaminated a cow pasture adjacent to the Windscale reactor site. The leak was detected by a radiation monitor in the ventilation stack of the reactor and the latter was scrammed. Radiation levels in the pasture and in cow milk subsided below natural background levels in a few months.

Besides in-core fuel element cracking, other typical accidents of a nuclear nature have been the accidental dropping of a spent fuel element which was pulled out of the reactor core for transient storage in a swimming-pool before shipment to a reprocessing plant (see Section 5.2). Dropping a fuel element can sometimes crack the cladding or fuel rod, resulting in release of fission product gases (mostly xenon-133 and krypton-85). Other nuclear accidents which are not strictly reactor related but which happened in fuel processing plants (Section 5.2.2) are the accidental spilling or leakage of radioactive fission product waste from temporary or permanent storage tanks. So far, no radioactive leaks or spills have ever caused a fatality due to radiation, and all have been remedied.

Both the US and Russia have had accidents with nuclear-powered submarines. The US lost the 'Thresher' on April 10, 1963 in the North Atlantic ocean, and the 'Scorpion' in May, 1968 near the Azores. Russia lost the 'Kursk' on August 12, 2000 in the Barents sea. These accidents did not involve nuclear supercriticality malfuctions, but were due to non-nuclear operations mishaps. The US Thresher and Scorpion are believed to have sunk because of excessive overpressure or collision at great depth in (experimental) dives, while Russia's Kursk sank when an undersea launch of one of its torpedos misfired and exploded. In all cases the entire crew of 129, 99, and 118 men lost their lives. Reactors on the sunken subs shut themselves down automatically as designed. If not salvaged, the reactors will stay subcritical for ever and pose no danger, contrary to alarmist's claims.

6.6. NUCLEAR CRITICALITY ACCIDENTS

Nuclear accidents are most serious if they involve uncontrolled supercritical masses of fissile material (see Subchapter 5.1). Since the beginning (1945) of the age of nuclear power, several purely nuclear 'criticality accidents' have taken place. It is important we diagnose how they happened, so repetitions can be avoided. We first briefly describe each accident and then comment on them.

During the ultra-secret Manhattan project, two criticality incidents occurred at Los Alamos, New Mexico. One was in August 1945 and the other in May 1946, each causing one death. In both cases, too much dissolved U-235 or Pu-239 were poured in a container, causing fission and a supercritical 'flash' with release of lethal amounts of non-shielded radiation (more than 500 rads). In June 1952, an experimental Chicago reactor overexposed four workers (maximum dose was 176 rad) before it was shut down; all workers fully recovered. On June 16, 1958, a supercritical mass of U-235 was accidentally formed in a drum at the Y-12 fuel processing plant in Oak Ridge, Tennessee. The super-critical U-235-bearing solution 'flashed', overexposing eight employees to radiation (20–230 rad). All men were hospitalized but survived. Three were released in ten days, and the other five on July 20, 1958. Still another criticality accident occurred at a Los Alamos plutonium processing plant on December 30, 1958. A supercritical mass of plutonium was created in a receptacle due to a piping connection error, and flashed. One man died in this accident one and a half days after the accident.

An experimental US Army reactor called the SL-1 went accidentally supercritical on January 3, 1961 at the US Reactor Testing Station in Idaho, when careless experimenters pulled out all control rods. It caused a core meltdown and explosion that killed the three experimenters standing on the bridge above the core. Two died instantly due to flying debris from steam-generated overpressures while the third man died hours later from combined concussions and radiation. Violation of safety rules and unfamiliarity with reactor physics were blamed for the accident. In summary, criticality accidents killed six workers in the US in 1945–1961: three in fuel-handling and three in reactor operations.

In Vinca, Yugoslavia on October 15, 1958, researchers were experimenting with a nuclear reactor when the core went supercritical, causing six people to receive serious overdoses of radiation. One died shortly after the accident, but the other five recovered after receiving bone marrow transplant injections at the Curie Foundation's Institute of Atomic Hygiene and Radiopathology in Paris.

On September 30, 1999, three workers at a nuclear fuel reprocessing plant operated by JCO Company, Ltd at Tokaimura, Japan, mixed batches of uranyl nitrate solutions in stainless steel buckets with 18.8% U-235 enrichment. The total amount of uranium was not to exceed 2.4 kg to prevent criticality, but being in a rush, they added a total of about 16 kg into a precipitation vessel. This amount went critical and started to boil, setting off radiation monitoring alarms some distance away on the site (there were none in the workshed with the precipitation vessel). Upon hearing the alarms and seeing a blue glow in the vessel, the workers ran off. Two were found to have received an estimated 1700 rem and 1000 rem, and a third one 300 rem of radiation. The first two men, whose white-blood-cell counts dropped to zero, died a few weeks later. Bone-marrow transplants were to no avail. The apparently self-controlling critical solution in the precipitation vessel continued fissioning and boiling for 18 hours before it was terminated by draining water from a cooling jacket around the bottom half of the vessel which had acted as a neutron moderating reflector. The JCO workers had not had any nuclear training and were unfamiliar with the concept of nuclear criticality. They had previously worked with uranyl nitrate batches that were only 3% U-235-enriched for use in power reactors, and the 18.8% enriched material was a special job for a research reactor order. They were unaware that a higher enrichment was much more serious in regard to staying within prescribed quantity limits. After this accident, many new government rules and regulations have been introduced at all of Japan's nuclear reprocessing facilities, including mandatory nuclear criticality physics instructions for all workers handling uranium.

Before describing the Three-Mile-Island (TMI) and Chernobyl reactor meltdowns, we mention that some nuclear accidents took place in Russia during the 1950's and 1960's, which were kept secret by the Soviet government at the time. After the cold war ended, Russian scientists reported that one nuclear accident occurred at Russia's plutonium production facility in the Urals, when too much fissionable material (a supercritical mass) was pumped into a storage tank for liquid 'waste' uranium solutions. The supercritical liquid mixture started to boil violently and blew radioactive material all over the site. After dilution of what was left and temporary transfer to other tanks, the radioactive liquid solutions were later dumped in an artificial lake which still emits low levels of radiation today.

The Three-Mile-Island (TMI) accident in Pennsylvania took place on March 28, 1979. One of the two operating PWR reactors at TMI developed a failure of a water pump in the secondary coolant loop that takes heat from the primary coolant to the steam turbines (Briefs 9 and 10). This pump failure initiated a reactor scram that halted the fission chain reaction, but decay heat kept heating the water in the primary pressurized water loop. Pressure build-up in this loop then caused a relief valve to open which allowed some primary coolant water to convert to steam and exit. Unbeknownst to the reactor operators the relief valve was stuck and stayed open, allowing more and more water/steam to escape from the reactor core and causing the latter to overheat. All the while the reactor instruments did not detect that the relief valve was stuck, and falsely indicated that the core was covered with high-pressure cooling water, while in fact the core was partially bare and the high pressure was due to an increasing buildup of steam. The operators mistakenly believed that the core was getting too much pressurized water and shut off the emergency core cooling system (ECCS). This action exacerbated the situation and the overheated core started to melt. Radiation monitors sounded alarms indicating that the core had released vaporized fission products. It finally became clear that the worst had happened: the core had disintegrated and suffered a (partial) meltdown. The containment vessel did perform its function however and kept all nuclear debris contained, as the core slowly cooled down. Two days later, a concern arose that hydrogen gas (H_2) might be building up inside the containment vessel due to radiolysis of water (H_2O) by slowly decaying radioactive products. This turned out to be a false alarm based on erroneous interpretation of instrument data. Release of some steam into the atmosphere which carried minute amounts of radioactivity, exposed those living near the reactor to a dose of at most 1 mr, less than one hundredth from natural sources (see Brief 17). While very costly to the utility, the

TMI reactor meltdown did demonstrate the basic soundness of TMI's reactor design and safety features, particularly its containment vessel. Not one person received a radiation overdose in this 'maximum credible accident' (= reactor meltdown).

The most disastrous nuclear reactor accident in the history of nuclear power took place on April 26, 1986 at Chernobyl in the Ukraine. The following is from a report of a special meeting held by the International Atomic Energy Agency on August 25–29, 1986 in Vienna, Austria [Ref. 32]. The Chernobyl-4 reactor was a graphite-moderated boiling water model of Russian design labeled RBMK-1000 (see Brief 23). Tests were being conducted at Chernobyl-4 to check out a new voltage regulation system that allows the inertia of the steam turbines during reactor shutdown to continue delivering enough power during the run-down to pump primary water coolant for a while, before the emergency core cooling system (ECCS) would kick in and take over. The tests involved the powering down of the reactor and the temporary shut-off of some of the core cooling pumps. After several interruptions of the testing program because of electric power grid demands from other units, the power-down was finally carried out. But it went far below the normal 1000 MW(e) operating level to 10 MW(e) instead of the intended 300 MW(e) level. The operators belatedly corrected for this overshoot by pulling the control rods out again. But because of the usual xenon-poison build-up common to all reactors after a (partial) shut-down, the reactor behaved sluggishly. To speed up a return to the desired 300 MW(e) level, they moved the control rods back out excessively and in addition pulled out some manually operated emergency standby control rods. This caused the reactor's reactivity margin to drop below safe limits in violation of operating procedures. Normally this would have initiated an automatic reactor scram, but the operators had turned off this electronic safety feature to speed up their tests. The reactor slowly regained power at first, but cooling water was colder and flowing at higher rates than usual. This caused a decrease in the formation of steam that ran the turbine which was being tested (the other one had been shut down). To counter the excessive water coolant flow, the operators turned off the emergency core cooling system (ECCS) which they were afraid might automatically kick in and exacerbate the steam deficiency that their instruments indicated. When more steam was finally produced, reactor power suddenly started running up exponentially (a 'prompt critical') which the control rods were unable to compensate for fast enough because they were overly withdrawn beyond the margin of safety, and the automatic scram control circuit had been turned off. In a graphite-moderated water-cooled reactor the coefficient of reactivity becomes *positive* when too much water turns into steam inside the core. This means the neutron multiplication chain reaction is enhanced when the temperature increases due to increased replacement of liquid water with steam. Had the reactor been a water-moderated and -cooled PWR or BWR (see Sec 5.1.1) both of which have *negative* reactivity coefficients, the reactor would have shut itself down in this situation, even with all control rods out and no water flow.

Chart records retrieved from the reactor operating room showed that the following chain of events took place just before the Chernobyl-4 reactor disintegrated. When the ractor operator saw the exponential rise of power, he pushed the emergency manual scram button that activates the dropping of all control rods back into the reactor core. A few ominous shocks were heard next and when the rods appeared not to reach their lower stops, the operator deactivated the electromagnets of the rods to allow them to fall by gravity. A few seconds later two explosions were heard, one shortly after the other, and burning lumps of material shot into the air from the reactor building, some falling on the turbine hall and setting it on fire. Two of the three reactor operating technicians were killed instantly by flying debris, while the third one died an hour later.

Since the Chernobyl reactor had no overall containment vessel and was housed in a hangar-like enclosure, air immediately penetrated the cracked reactor vessel and started to burn the overheated graphite moderator in the reactor core. This fire evaporated large quantities of radioactive fission products in the core and carried them into the air. The resulting cloud of radioactive debris deposited fall-out contamination on the bedroom community of Pripyat, 10 km away. Firemen from nearby Chernobyl and Pripyat were called and rushed over to extinguish the reactor fire, but none had ever been briefed or were aware of the effects of nuclear radiation overexposures. A total of 28 of these firemen and other rescue workers died later over a period of three months due to heavy overdoses of radiation. Three more died from heart attacks and 11 succumbed from attributable medical complications one to ten years after the accident. Another 30 rescue workers suffered permanent disabilities. The final human casualty score was 45 dead and 30 permanently disabled.

The local administrative bureaucracy reacted slowly and did not immediately grasp the significance of nuclear fall-out radiation. Only after radiation monitoring service personnel from the Chernobyl plant detected serious levels of radioactive fall-out radiation at Pripyat, was the entire population of the town avacuated. The 45,000 residents were also given potassium iodide (KI) pills to prevent possible thyroid take-up of radio-iodine from the fall-out. This proved effective and caused no side-effects. Pripyat is still a ghost-town, but none of its former inhabitants suffered permanent health problems attributable to the Chernobyl-4 fall-out. Alarmists claim that thousands may still die in the future from fall-out consequences, but there is no proven scientific basis for this.

Fall-out also fell on more distant meadows where cows were grazing. Milk from these cows was later found to be contaminated with radioactive iodine from the fall-out. A dozen or so children who had consumed this milk were subsequently found by doctors to have some radio-active iodine in their thyroids and were hospitalized for a while. None died from the exposure and all were dismissed after most of the radioactive iodine had decayed away. The whole affair could have been prevented if more immediate action had been taken to quarantine all lifestock grazing or feeding on land under the path of the fall-out cloud, and if milk and other agricultural products from the effected area had been monitored for possible radioactive contamination. Also more potassium iodide (KI) pills should have been

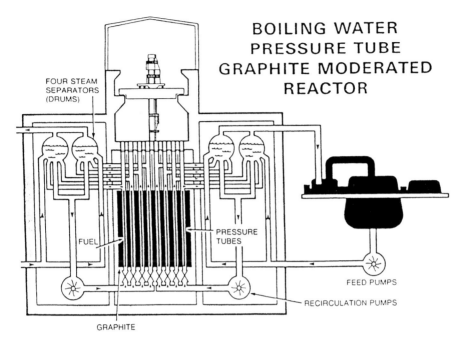

BRIEF 23. DIAGRAM OF THE CHERNOBYL RBMK-1000 REACTOR

given immediately to all people living under/along the path of the atmospheric fall-out cloud over at least a distance of 60 km.

The Chernobyl disaster has caused Russian reactor designers to completely overhaul and substantially improve reactor safety features. Prior to Chernobyl these had been deemed inadequate and insufficient by US and WestEuropean standards. A Chernobyl-like reactor accident with atmospheric release of fission products could never happen with US or West-European reactors, because of the insistence that an overall containment vessel surround all reactor activities so it can keep all nuclear debris contained in the event of a 'maximum credible accident' (MCA). An MCA includes unexpected earthquakes of magnitude 8, the crash of an airplane on the reactor, etc. Although a reactor core and moderator is itself confined by a reactor pressure vessel that can withstand limited overloads of heat and pressure, the containment vessel surrounding the reactor vessel and gamma shield is the final barrier beween nuclear reaction products and the atmosphere in case the reactor vessel disintegrates. The TMI accident was of course unintentional but demonstrated the basic soundness of the original 1970's reactor safety design features. Not one person was hurt. Since the TMI accident, a large number of improvements in reactor operational safety and reactor design have been made, making it virtually fool-proof against human error. Unpredictable earthquakes and airplane crashes will cause automatic shut-down scrams, and as mentioned, even if all control rods froze instantly, reactors with negative reactivity coefficients shut themselves down.

Sabotage or suicide missions by terrorists could damage a reactor, but they would be unable to initiate a Chernobyl-like explosion in a modern reactor. A negative reactivity coefficient makes it physically impossible for reactors to go much beyond the onset of 'a prompt critical' and shuts them down automatically if too much fission heat is generated.

The International Nuclear Event Scale (INES) ascribes a value of 4 for the Tokaimura incident, a 5 for the Three-Mile-Island accident, and a 7 for the Chernobyl disaster. This scale is exponential like the Richter scale for earthquakes, so that a 5 is ten times more serious than a 4, and a 7 is a thousand times worse than a 4. Even though 45 fatalities at the worst nuclear accident in Chernobyl are deplorable, sixty years of worldwide nuclear electric power operations with less than a hundred human casualties is remarkable when compared with the thousands of people that have perished in airplane crashes and refinery or chemical industry fires. It should give one great confidence that the nuclear power industry can be and is a safe human enterprise (and a whole lot safer than warfare!).

CHAPTER 7

MEASURES AND MEANS TO CONTROL THE GLOBAL USE OF NUCLEAR WEAPONS

7.1. THE NUCLEAR AGE AND WORLD REALITIES

Because uranium and plutonium are also sources for nuclear weapons, a lot of public concern has been expressed about the danger these materials might pose if they were clandestinely diverted from use in civilian nuclear power plants to more sinister applications. These concerns are the true motives behind many who want to abolish nuclear power. They fear that proliferation of nuclear weapons is promoted by the existence of nuclear power reactors. Thus some keep fabricating unsubstantiated charges that there are 'problems' with nuclear waste disposal (e.g. at Yucca mountain) while their real intention is to disrupt development of nuclear power. To be believable, they should isolate and separate their valid concern for nuclear weapons proliferation from the issue of whether the operation of nuclear power plants and disposal of radio-waste can be done safely. The two issues are totally distinct and different. In this book I hope I have clearly shown that the second issue is a non-issue and that we have no choice but to expand nuclear electricity production if we want to avoid a no-oil calamity in 2030. Regarding the first issue, some thoughts on promoting non-proliferation are presented below. They require a review of non-scientific unpredictable human mass behaviour.

The question is how one can prevent construction and use of new nuclear weapons in addition to those already possessed by the USA, Russia, Great Brittain, France, China, India, Pakistan, Israel, South-Africa, North-Korea, and perhaps others (South-Africa destroyed its small nuclear stockpile around 1993). Would a world-wide ban or moratorium on (additional) nuclear power plant construction as some suggest, promote this goal? In anticipation of arguments I shall give below, the short answer to this question is an emphatic NO! Just the opposite effect will be promoted: increased nuclear weapons acquisitions and proliferation. I base this belief on history and world realities.

Uranium does exist on planet Earth just as surely as the raw materials for nitroglycerine and dynamite. Nobody can change this unalterable fact. Knowledge of how to extract explosive energy from uranium or nitroglycerine for destructive uses is also universal. No amount of book-burning or suppression of dissemination of scientific information is going to change that. Even if one was able to eradicate knowledge of uranium fission in this generation, the next generation will rediscover it for sure. If the present generation shirks its responsibility and fails to organize a world where peaceful use of nuclear energy (Eisenhower's vision) can thrive, it will only pass the job on to the next generation who will try to resurrect it amidst enhanced energy shortages, economic upheavals, and probably widespread warfare for control of the last remaining fossil fuel and coal reserves.

Some anti-nuclear activists who never built an electrical, optical, or mechanical device in their life, often brandish the terms 'nuclear technology' and 'nuclear materials' without knowing precisely what they mean. They seem to think that the adjective 'nuclear' somehow permits separation of techniques and materials used in making nuclear weapons from other technologies and materials, so they conveniently advocate banning all 'nuclear technology and materials' with the stroke of a pen. The making of a nuclear weapon entails using nuts, bolts, special metals, electronics, chemicals, etc, etc, which are the same as used in hundreds of modern equipments in laboratories, aviation, the space program, mining, chemical industries, and many other endeavours. The hardware technologies (calutrons, centrifuges, e-beams, lasers) used in separating medical isotopes for nuclear medicine are the same as in isotope enrichment of uranium. For diagnostics, every analytic laboratory in the world utilizes the same 'technologies' as in uranium enrichment. One cannot simply issue an edict prohibiting the use of 'nuclear technology' without incapacitating nuclear medicine, biotechnology research and development, and many other high-tech operations. The only difference between isotope separation of uranium and other isotopes is the scale, i.e. the quantity that is processed. For nuclear power plants (and weapons), one needs kilograms to tons of material, whereas in most other isotope separation applications only milligrams to grams are needed.

Placing export restrictions on technical equipment to countries that might want to develop nuclear power (or possibly weapons), only causes such countries to develop their own infrastructure for building 'nuclear' components, as Pakistan did. Manufacture of high-tech hardware is a matter of investment capital and collecting capable engineers and scientists (many are unemployed and available). It does not require any 'secret' knowledge. The USA once held nuclear fission 'secrets' exclusively for a few years during and after WW-II. Some people seem to be under the illusion that only the USA possesses 'the nuclear secrets'. Secrecy surrounding the US nuclear labs at Los Alamos and Livermore only involves engineering design details, test results, locations, and hard-won quantitative data about nuclear weapons, which of course cannot and should not be disclosed to the public. But it should not give one the false impression that no one else can make nuclear weapons.

It is more productive to pursuade nations to sign the Non-Proliferation Treaty (NPT) monitored by the International Atomic Energy Agency (IAEA) than to boycot them (Subchapter 7.3). Rather than encumbering international trade and promoting enormous bureaucracies and paperwork that invites corruption, the IAEA should place resident monitors/inspectors (rotated yearly) in signed-up NPT countries which have declared possession or planned possession of complete uranium fuel-cycle facilities. Only countries that refuse to sign the NPT should perhaps be ostracized and kept under continuous pressure to join the world federation of NPT nations.

The problem of a regime change from peaceful to hostile, which might alter or rescind an earlier NPT agreement (e.g. North Korea), is always present and must be dealt with through the United Nations. An unlikely but possible ultimate solution is that all nuclear fuel-cycle operations in the world be placed under management of the IAEA. That is, all fuel-cycle operators in the USA and Europe like USEC, URENCO, AREVA, and all such operations in Russia, China, Japan, India, Pakistan, etc. could be controlled and operated by one global organization. However without competition, innovations and productivity may become stifled and cost-fixing may become a problem. Also, it is improbable that the military would allow their direct access to be severed. A more realistic ultimate solution might be co-management, in which employees of the IAEA NPT-monitoring staff are interspersed among the top administrators of uranium fuel handling organizations.

Some realities in today's world with relevance to nuclear weapons and warfare between nations (terrorists will be considered later) are the following:

1. Many nations and ethnic groups fear domination by neighbours which they resent. (Examples: Israel vs Palestine; Iran vs Iraq; India vs Pakistan; North-Korea vs South-Korea; Catholic Irish vs Protestant Irish; North-African Arabs vs African Blacks; Christian Serbs vs Moslem Bosnians and Albanians; Greek Cypriots vs Turkish Cypriots).
2. No country (least of all the USA) wants to have to rely on another country for support in a conflict that threatens its existence.
3. No country wants to accept the idea that the USA will be the super boss of the world and dictates world policy for everyone.
4. No country believes that the United Nations can settle all wars and disputes fairly.
5. Any nation that presently has abundant energy resources whether oil, coal, natgas, or uranium can wield power. (Example: the OPEC countries).
6. Any country that produces large amounts of electric power efficiently and runs modern industries economically, can produce competitive products that allow it to be in a powerful position opposite other nations. (Examples: Japan, Europe, USA, China).
7. Presently, fossil fuels energize all transportation fleets in the world. Their supply is controlled by a small group of giant oil companies and OPEC nations. After oil is gone, a shift to uranium as the prime energy source for producing synfuels will assuredly take place. Expansion of nuclear power cannot be stopped by opponents simply because they fear it would lead to nuclear weapons

proliferation. Countries that possess fuel-cycle capabilities (Subchapter 5.2) and have access to uranium ore, will become the future suppliers of portable synfuels, like the big oil companies of today. It should be no surprise that countries like Iran also want to develop such a capability before their oil reserves are depleted. Calls for international sanctions by present uranium-cycle countries to keep others from joining their club, will be perceived as attempts to protect present monopolies.

8. Any sovereign nation of moderate size and means can design and build nuclear power reactors and enrich and reprocess uranium if it decides to do so. The USA, Russia, Europe, China, Japan, India, Pakistan, South-Africa, or Israel have no special privileged information except that they are ahead with research, development, and improvements of the fuel cycle, and have spent at least three years and billions of dollars on it.

9. Any sovereign nation of moderate size, if it so desires, can build a nuclear bomb by producing plutonium in a breeder reactor, or by high-level enrichment of uranium. It takes only 3 to 4 years from the moment such a decision is made, to possess a weapon. A country might decide to do so if it feels threatened. It will intensify its efforts to develop the technology even more, if the super-powers prohibit equipment exports to it (e.g. Pakistan, North-Korea, Libya).

10. Nations in the nuclear weapons club plan to use their nuclear weapons only if another nation threatens to overrun them. The best protection against nuclear blackmail is to possess a nuclear retaliation capability. (Examples: USA and former USSR; India and Pakistan).

11. Detonation of one or even ten nuclear weapons, although abhorrent, does not mean the end of the world.

12. Many poorer countries have the perception (right or wrong) that citizens of the prosperous USA and other big powers flaunt attitudes such as: (a) 'We know what is best for you'; (b) 'If I am nice to you, you should be nice to me'; (c) 'If you are bad, we are not going to help you'; (d) 'We have all the right answers'; (e) 'We have all the brains, money, and know-how; we don't need your help'.

I listed these causes of discontent and insecurity in the world as a reminder that humanity is not at peace, and at any time can generate conflict and warfare. In the 21st century, such hostilities may involve the use, or threatened use, of nuclear weapons. This in turn has lead many pacifists to oppose development of nuclear power even for the peaceful generation of electricity. But to curtail nuclear power programs, hoping to prevent nuclear weapons proliferation will not change realities (1) through (12.) Halting nuclear electricity generation would be tantamount to committing national suicide. Thanks to earlier forward-thinking governments and industries worldwide, nuclear electric power generation was solidly developed during the 1950–1990 period. Should governments now accede to the whims of special interest groups who clamor for the abandonment of nuclear power, all earlier achievements by diligent energy planners and engineers would be lost for no good reason. If the US abolished nuclear electricity, escalating costs of imported oil, synthetic fuels, and purified coal would assuredly lower the standard of living and impoverish us as we approach 2030 (before total oil depletion arrives). The US would be unable to compete in international markets, while more foreward-thinking disciplined countries like China take over leadership positions in nuclear power expansion. The proverbial 'terror of democracy', in which a majority of misinformed non-professionals rules and ignores the advice of seasoned professionals, may come to haunt us.

The notion that solar- and wind-power could replace nuclear fission power as a substitute for oil, is incompatible with reality as shown in Chapters 2 and 3. At best it might provide up to 10% of man's energy needs after the year 2030, when oil shortages will become severe[10]. That one could learn to do without another 90% through energy saving techniques (insulation, etc.) is also a pipedream. The other 90% are needed to make synfuels and run the heavy industries that make our cars, trucks, airplanes, ships, bridges, etc. One cannot melt and forge steel or aluminum without enormous amounts of electricity or heat energy, which windmills and solar cells alone just cannot provide. No amount of wishful thinking can change the basic laws of physics and economics. I also don't believe that back-to-nature protagonists will be able to convince third-world citizens they should stick to tending camels and llamas, and burn animal dung for cooking food and heating their huts. Most young people in the third world want energy-consuming cars, stereos, televisions, computers, refrigerators, etc.

Though I digressed from the subject of nuclear weapons proliferation, the main point here is that the privileged in the world cannot force others on the planet to accept their philosophies and viewpoints. One can influence, frustrate, and delay the realization of other people's aspirations but one can never control them forever, as shown by the collapse of totalitarian communism. Should the USA impose a nuclear power moratorium like Germany did, other nations who comprehend the gravity of the upcoming oil-depletion crisis more clearly, will take over nuclear power development. This includes the capability of building nuclear weapons. Acquiring or stockpiling nuclear weapons will be encouraged particularly if it appears that the balance of power is shifting in the world. If a country is no longer sure of its own security, it is going to take countermeasures. No amount of pleas by pacifists and others with noble intentions are going to change such a trend, as history has shown over, and over, and over again. In conclusion, it is my belief that the result of any moratorium on further development of nuclear power will enhance worldwide nuclear weapons proliferation instead of abating it, as illustrated in Brief 24. How we might minimize and best cope with nuclear weapons proliferation is the subject of Subchapter 7.3, after we first consider the problem of terrorists who wish to blackmail the world with nuclear weapons.

---

[10] Recent announcements that Denmark obtains 20% of its electricity from wind-mills, only applies to that country's electric grid. If energy from natural gas and petrol used by its transportation fleets are added, the wind-energy percentage of *total* energy usage is less than 10%

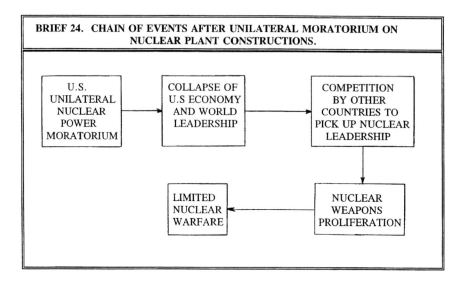

BRIEF 24. CHAIN OF EVENTS AFTER UNILATERAL MORATORIUM ON NUCLEAR PLANT CONSTRUCTIONS.

7.2. SAFEGUARDS AGAINST TERRORISTS

It has been suggested that a terrorist gang could steal nuclear fuel elements from a reactor and make a bomb out of it. As already mentioned in Section 5.1.3, the fuel elements of a nuclear reactor are totally unsuitable for a bomb. If terrorists wish to build a nuclear weapon from reactor fuel elements by stealing them from a nuclear plant, they must steal several hundred highly radioactive fuel elements from two or more nuclear reactors. They must first construct a fifty-million-dollar special fuel reprocessing plant whereto they can take the stolen reactor fuel. Next they would have to spend six months or so to carefully extract fissionable uranium and plutonium in such a plant and figure out how to construct a nuclear bomb, which requires a technically very advanced group of engineers. Clearly from the review of criticality accidents in Section 6.6, ignition of a nuclear fissionable critical mass (e.g. Tokaimura) does not produce a Hiroshima-like explosion. To make an effective nuclear bomb requires extensive research and backing by a sovereign government. For an illegitimate gang trying to pull off such a project clandestinely after armed robbery of two or more reactors, is certain to catch the attention of the world's intelligence agencies and international police (interpol).

Another possible target for terrorists might be 'hot spots' in the nuclear fuel cycle (Section 5.2), where large amounts of enriched uranium and plutonium occur in concentrated form, namely enrichment plants, reprocessing plants, and nuclear fuel element fabrication plants. Even after overpowering all the guards of such heavily guarded installations, and carrying off all the raw plutonium or highly enriched uranium, the job of making a workable nuclear bomb from these materials is still a very complicated task. Actually many precautions are taken at enrichment, reprocessing, and fuel fabrication plants, that make it very difficult to obtain enough material for a bomb in one raid. To avoid any accumulation of a critical mass, many of the operations in these plants are compartmentalized so that there is never more than half a critical mass in any one section. To obtain enough for a nuclear weapon, one would have to raid many compartments, each one of which can be hermetically sealed in case of an emergency or alarm from the guards. If nevertheless succesful, terrorists would have to take precautions to keep sub-critical quantities of nuclear fuel taken from different cells separated, lest they don't incapacitate themselves with a supercritical flash (see Section 6.6). If on a suicide mission they might do this intentionally, but the resulting nuclear flash and resulting damage would remain localized, similar to the Tokaimura criticality accident in Japan. There would be a minimal effect to the general public.

Stealing small amounts of U-235 or Pu-239 by a corrupted insider in any one of the three fuel handling plants of the fuel cycle is easier done but still difficult. Small processing losses of fuel always occur and it is possible that if such stolen losses are small enough, they would go undetected. Modern accurate accountability controls and certifications, thorough personnel screening for security clearances, and monitoring of workers entering and leaving the plant at all times, should minimize the chance of small-quantity thefts of fissionable materials. From many years of operation, the statistics and amounts of processing losses in these plants are well established now and discrepancies can be quickly detected. If someone stole such small amounts that losses would be undetected, he would have to spend many years collecting before he had enough for a critical mass. Quite possibly he would have changed his mind by then to make a nuclear bomb.

Still another operation that might be vulnerable to terrorist attacks is the shipment of unirradiated (fresh) U-235 or plutonium fuel elements when they are transported from the fuel fabricator to a reactor plant. To minimize the consequences of armed attacks and thefts, precautions are usually taken which include insuring that the amount of fuel is less than required for a critical assembly. Thus the theft of fuel from two or three transports would have to be carried out if the objective of the terrorists is to obtain enough for a nuclear weapon. Should a shipment of irradiated fuel elements with freshly bred plutonium be attacked and robbed on its way to a reprocessing facility, not much could be done with the stolen fuel since it is intermixed with Pu-240 and other neutron 'poisons' that prevent nuclear chain reactions. It is the reason why the shipment was on its way to be 'reprocessed' to take out undesirable elements that inhibit nuclear fission. Thus unless the terrorists possessed a reprocessing facility or just wanted to produce a scare, such a heist would be useless to them.

Recently the news-media have disseminated a lot of 'chatter' about 'dirty bombs'. Such bombs would be made of ordinary explosives except that they would be seeded with radioactive materials, most likely fission products. They rely mostly for their effectiveness on the fear-factor of many citizens for things 'nuclear'. While a bomb is still a (lethal) bomb, the fact is that little extra is accomplished in terms of casualties by putting radioactive elements in a bomb charge. Except for casualties killed by the non-nuclear shockwave of the explosion, bystanders covered with radioactive dust will survive and can be de-contaminated by washing off the dust and/or taking a shower. Should radioactive elements from the bomb blast include iodine, strontium, and cesium, and should some dust have been inhaled or ingested, pills can be taken by victims to help eliminate these biologically unfriendly species (see Sections 5.3.2 and 6.6). Radioactive dust is easily traced with radiation monitors. Objects covered by it can be decontaminated with 'radwaste', a soapy solution that soaks up most radioisotopes. In conclusion, 'dirty bombs' can never have the same effect as a real nuclear bomb explosion with kilometers of massive destruction; damage would be the same as from a non-nuclear device except it generates fear and requires a radioactivity decontamination crew to clean up.

Instead of stealing fuel elements out of nuclear reactors, or obtaining nuclear fuel from enrichment and reprocessing/fabrication plants, or from transports between them, a more likely possibility is that a terrorist gang might attempt to steal a complete nuclear weapon from any one of the nuclear arsenals around the world. Though contents and locations of such arsenals are kept secret, a well-organized terrorist gang might infiltrate the military and discover how to break into one. The military have of course considered this and have taken appropriate countermeasures. The theft of a complete anti-aircraft missile some years ago in Germany shows that stealing a military weapon is not entirely impossible. One special measure for nuclear weapons that prevents or frustrates such a possibility is that nuclear warheads are usually composed of two or more separate parts that can be assembled only by one or two officers to produce a trigger-ready critical mass. These officers must have special tools to 'arm' a weapon but can only do so if given coded instructions. While details must of course remain secret, it is believed weapon components that can be mated are stored at different locations only known to and accessible by the special-duty officers, whose identities are kept confidential and who are frequently rotated. Thus should a terrorist succeed in breaking into and entering a weapons storage bunker, he would still only be able to get (less than) half the critical amount needed for a fully operational nuclear bomb. If tampered with by non-coded tools, alarms may go off and neutron poisons released internally that make the component unfissionable, even if another stolen bomb half (or part) were to be brought in contact with it to increase critical mass.

Military and professional nuclear engineers around the world have worked with nuclear materials for over sixty years and have long learned how best to safeguard these materials against unfriendly intervenors, including terrorists. Besides some of the precautions already discussed, many additional safeguards exist (some kept secret) which so far have succesfully prevented any terrorist from capturing a nuclear device. Unless a rogue sovereign nation cooperates, clandestine capture of a functional nuclear weapon by terrorists is about as impossible as breaking into Fort Knox to steal gold. A much simpler (and more frightening) way for extremists to terrorize a civilian population is their use of chemical or biochemical agents.

After the September 11, 2001 attacks on the New York World Trade Center, security has been considerably increased at power plants, water treatment facilities, chemical plants, oil refineries, and other vulnerable factories. Armed guards, monitors, and concrete barriers have been placed around most of them to protect against terrorist attacks. The question often asked is what havoc can result if in spite of all these measures, a terrorist suicide team would succeed in entering a nuclear power plant after killing the guards and capturing the reactor operator(s). Most likely, the reactor operator would have scrammed (shut down) the reactor before they could get to him. This makes it physically impossible for the reactor to restart immediately due to so-called 'xenon poisoning' (built-up of neutron absorbing xenon from post-operation nuclear decay). Should terrorists capture the operator somehow before a scram, and force him at gunpoint to pull out all control rods, the reactor would still shut itself down because of the negative reactivity produced by overheating. It also automatically activates the emergency core cooling system (ECCS). If the terrorists disabled the ECCS, the core would partially melt. However except for some gaseous xenon and krypton, core contents would stay in the core vessel (Subchapter 5.1), because core materials are solid or (now) a liquid melt. The result would be the same as what happened in the TMI accident (Subchapter 6.6). The large outer containment vessel (Brief 9), would still keep core-vessel-escaped gases contained should the core vessel crack.

In a suicide mission, terrorists would probably detonate a bomb in the reactor control room, once inside. This would destroy all reactor controls, but as stated, the reactor shuts down automatically. It is almost impossible for bomb fragments to penetrate the 2-inch (5 cm) thick steel shell of a contaiment vessel. Even armor-piercing mortar shells would have difficulty puncturing it. Should it be breached, the negative pressure maintained in most containment vessels would inhibit outflow of radioactive core gases, and only permit inflow of air (or pure nitrogen). Unlike the Chernobyl reactor whose core was made of ignitable graphite and which had no containment vessel, core-burning fires cannot break out in a water-moderated reactor. The radioactive cloud produced at Chernobyl could never occur. In conclusion, a succesful terrorist infiltration of a civilian nuclear power plant would at worst result in a TMI-like core melt-down, with reactor operations personnel most likely killed.

7.3. THE NON-PROLIFERATION TREATY (NPT) AND INTERNATIONAL ATOMIC ENERGY AGENCY (IAEA)

With the unpleasant picture painted in Subchapter 7.1, one asks how one can best prevent or minimize the proliferation and use of nuclear weapons by unfriendly sovereign nations. We have already argued that inhibiting the expansion of nuclear electric power will set the stage for an enormous world energy crisis when oil runs out. Only with nuclear electricity is it feasible to produce sufficient future supplies of oil-replacing synfuels for our transportation fleets and enough power for our heavy industries, without global warming. By sharing man's accumulated knowledge and the planet's resources with everyone, more people in the third world will have hope for a tolerable existence, and chances are less that warfare will break out for control of the last remaining fossil fuels on earth.

A free exchange of nuclear reactor information world-wide, started under president Eisenhower's 'Atoms for Peace' program, would unite nuclear engineers and scientists from different nations and insures regular contacts. If somewhere in the world a country would suddenly place a nuclear project 'off limits', the world would be forewarned that someone may be 'diverting' for nuclear weapons. Pressure could then be brought upon that country to disclose its intentions and to cease further 'diversions'. If on the other hand information exchange about nuclear power technology were restricted between nations, a country could much easier conceal any work on nuclear weapons.

What still appears to be the best international solution to the nuclear weapons proliferation dilemma is the so-called Non-Proliferation-Treaty (NPT), and the NPT monitoring and enforcement program of the International Atomic Energy Agency (IAEA) headquarterd in Vienna under the auspices of the United Nations. The NPT is an international agreement signed by many (but unfortunately not all) countries in the world who pledge not to 'divert' uranium or plutonium fuel from use in nuclear reactors to use in nuclear weapons. By signing the NPT, countries also allow periodic and random inspections of their nuclear facilities by officers of the NPT enforcement arm of the IAEA. Any plant or facility in the nuclear fuel cycle that handles uranium or plutonium is defined to be a nuclear facility. Thus nuclear reactors, fuel element fabrication plants, reprocessing plants, enrichment plants, uranium mines, and radiowaste disposal facilities all qualify as nuclear facilities. Of these, the main 'hot spots' where diversion is most easily carried out are: (a) reprocessing plants; (b) enrichment plants; and (c) fuel element fabrication plants. The fuel placed in nuclear reactors is itself only 'reactor-grade' and unsuitable for weapons unless it is reprocessed. Even reactor fuel elements that use plutonium could not be changed into a bomb device without considerable reprocessing and reconstitution, requiring high-vacuum furnaces and other chemical processing equipment.

At the heart of a succesful nuclear safeguards program is a strict accountability program for uranium, thorium, and plutonium. Theoretically if one could track every gram of these three elements around the world, one would have a perfect NPT monitoring system. In the USA and other countries that signed the NPT, anyone possessing or overseeing work with enriched uranium or plutonium for whatever reason, must be licensed and by law must maintain an accounting of every gram of these materials. The accountability records must be open to inspection by IAEA agents at all times. While such accountability measures are enforceable in the USA and most other NPT countries, there are large gaps at present which makes a worldwide system fallible. Natural uranium can be bought freely from a number of sources in the world and can be enriched secretly in a non-NPT country. Even if present major uranium (and thorium) mines and suppliers would register and account for all their transactions, new ore deposits are discovered all the time and could be kept hidden.

Although the IAEA/NPT program is still imperfect, it is a positive step in the right direction. Instead of searching for alternatives (nuclear power moratoria, etc), it is the author's opinion that the only realistic approach is for all peace-loving nations to continually improve, strenghten, and upgrade the existing international nuclear safeguards program. One of the first objectives should be for all nations to join the IAEA/NPT program. Since many countries refuse to sign the NPT on grounds that it gives the superpowers and other non-signatories a big advantage in a future conflict (which is a reasonable position), one could perhaps provide for a provisional or modified NPT agreement. This agreement would allow the making of a limited number of nuclear weapons, provided they are registered by the IAEA, and with a promise they will never be used except in the defense of the country's existence or other restricted set of circumstances. It is better to know than not to know who has nuclear weapons. With that, a pledge is also made that the country agrees to the ultimate goal of worldwide total nuclear disarmament by all nations, including the present superpowers. To further pursuade a nation that presently refuses to sign the NPT, a guarantee of the country's sovereignty and boundaries signed by all other nations might be offered. This idea is admittedly simplistic and difficult to implement in for example the Middle East or the Koreas. However negotiations might at least be initiated and an attempt made to find possible solutions that avoid future armed conflicts.

Returning to the technical aspects of a NPT-governed future world, it is in principle possible today for a global satellite system to monitor all major constructions, excavations, and earth-moving operations on our planet and to keep tabs on for example uranium and thorium mining. In fact the characteristic emissions from radon gas which is almost always present near piles or tailings of uranium ore may be used to detect such activities. However once uranium and thorium are concentrated and refined, detection and tracing of its whereabouts becomes more difficult. Although it may be possible to devise remotely monitorable tagging systems, at present it appears that a two-part global nuclear fuel tracing and monitoring system would be needed in which satellites would keep a continuous global watch over new uranium/thorium mining operations, and movements from the mines are tracked separately by a sophisticated computerized tamper-proof accounting system.

Assuming all nations have signed the NPT or a modified NPT so that IAEA inspectors can monitor fissionable fuel movements in every country, an IAEA team can be dispatched to any spot on earth where a new unannounced uranium/thorium mining operation is discovered by the global space-watch satellite system. All production of uranium and thorium at mines in any part of the world is thus logged and entered into the global accountability system of the IAEA. Shipments of ore from mines to fluoridation plants or other uranium pre-processing plants are also monitored and accounted for at all entry and exit points much like money-balancing operations employed in world-wide banking.

Since uranium is radioactive and emits a characteristic gamma that can be sensed by a detector, engineers can think up many schemes that could make the nuclear accounting system semi-automatic. Of course random inspections must still be made by IAEA officials to ensure that installed uranium monitors are operating properly and not compromised. False and secretly hidden monitors whose locations are randomly switched now and then may be deployed to protect against possible fraud. To design, develop, and install a reasonably tamper-proof monitoring system for nuclear accountability will require some in-depth studies but appears feasible. Special protection and measures against terror gangs bent on making a nuclear weapon as considered in Subchapter 7.2, would of course be an integral part of the overall worldwide nuclear safeguard program. The IAEA's future NPT enforcement organization may thus consist of one or more central terminals located at strategic locations in the world which collect space satellite data and ground-based monitor readings from around the globe. A daily balancing of the books would be carried out by a computer, and if irregularities appear anywhere, a special team of NPT officers is dispatched to the location for a check-out.

Anonymous rewards should be given to anyone who reports a cheating event or a conspiracy to steal nuclear fuel if proven correct. That such awards are given should be widely publicized. Jail sentences for guilty parties should be stiff and internationally agreed upon to deter anyone from trying to cheat the system. The NPT police force should have direct liaisons with interpol and national police forces in the world as well as with international intelligence agencies. A concerted international cooperative manhunt could be started if and when it is found that nuclear material has been stolen by criminals.

Countries that joined on the provisional NPT program with a small stockpile of nuclear weapons, will register these with the IAEA, but the whereabouts could remain a military secret of the country. Pledges are obtained from such countries that adequate safeguards are maintained to prevent some madman in the military establishment of the country from starting a nuclear war. Information on safeguard techniques to prevent such a possibility, as practiced by the USA and Russia, should be made available to all nuclear club members. It is hoped that ultimately a world order can be achieved where all nuclear weapons are turned over to the IAEA and reprocessed for fuel in nuclear reactors.

In organizing the IAEA's NPT enforcement organization, great care must be taken that it can function semi-autonomously and not be subjected to partisan political pressures from powerful members of the United Nations. People in key positions of this organization should be chosen from scientific institutions all over the world. They should have impeccable reputations for honesty, have reached a certain level of maturity, and be moderate in their personal political convictions. Their appointment to office must be approved multinationally perhaps by a two-third majority vote in the UN. It is very important that IAEA overseeers of the NPT agreements be impartial yet undeceivable. Accusations that a country is 'diverting' made by another nation for political reasons, should be evaluated with total impartiality. No judgments can and must be made by the IAEA without hard technical evidence that the accusations are correct. It is important to keep in mind that any sovereign country has the right to possess and acquire nuclear reactors for electric power generation or radioisotope production, as well as all fuel-cycle processes of mining, isotope enrichment, fuel fabrication, reprocessing, and waste disposal. As long as an NPT-signatory country allows entry by IAEA inspectors to verify NPT compliance, one cannot demand that it must forego development of any fuel-cycle process only because it *might* lead to development of a nuclear weapon.

Some may still feel that the cost and labor involved in running and maintaining the IAEA's NPT program are excessive and that it will be ineffective. They still would like to simply ban all nuclear activities including nuclear power plant development. Such people should be reminded once more of the fact that the reason the NPT program is needed is because of the weapons aspects of nuclear materials, *not* nuclear electric power. We don't abandon flying airplanes because they can be used alternatively in warfare to bomb cities. Even if it was impossible to build nuclear power reactors from the same materials, the world needs a NPT program to curb or minimize chances of nuclear warfare. Knowledge that nuclear weapons exist and can be built is an irreversible fact today which we must learn to live with. When compared with the cost of military establishments of different countries, the international NPT police force is rather inexpensive, considering that such an organization can impede, moderate, and ultimately cause a ban on the use of nuclear weapons on our planet. It is worth every financial investment that is needed!

Those campaigning against nuclear electric power plants because of their fear of other Hiroshimas, are urged to study this book for the complete story ('the rest of the story' as US radio commentator Paul Harvey is fond to say). If convinced of the safety and pollution-free features of nuclear power plants and the benefits they bring to man, they ought to re-evaluate their thinking and campaign for a stronger and better IAEA-NPT police force; foreign governments that are still uncommitted should be pursuaded to join the NPT club. That nuclear energy will be interwoven with our future is a certainty whether wanted or not. Rather than fighting change as is the natural tendency of many, being open-minded and understanding of our changing world is more productive. These changes are less frightening when one comprehends the true issues and knows what needs to be done. If proper safeguards are developed, nuclear energy can become a blessing for all future generations instead of a perceived curse. The dire predictions made by some doomsday prophets of pending calamities due to energy and material shortages will only come true if the benefits from nuclear technology are withheld from our children.

CHAPTER 8

CONCLUSIONS, ACTION ITEMS, AND PREDICTIONS

Past and future large-scale exploitation of prime energy sources by man might be summarized as follows:

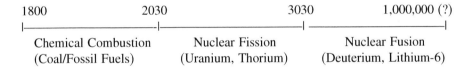

This time-line is based on the calculated resource depletion periods discussed in Chapter 3 and today's knowledge of atomic and nuclear physics. Most scientists are quite certain that present knowledge is sufficient to assess the order of magnitude of dormant energies in molecules, atoms, and nuclei. However how to best harvest that energy can change. Cold fusion (Subchapter 5.3) for example, if proven feasible, could alter the above timetable and bring fusion extraction schemes into the picture earlier. 'Dark energy' is another mysterious force of nature discovered by astrophysicists and cosmologists, which some day may be understood and perhaps utilized by man. However waiting for a miracle should not delay preparations now for the nuclear fixes needed by 2030.

To avoid global warming and a disastrous energy crisis by 2030, high-tech developments based on proven science must take place concurrently on four fronts:
 (I) Synfuel Testing, Mass-Production, Storage, Distribution, and Selection
 (II) New Propulsion Engines (Fuel Cells and Combustion Engines) – Testing and Mass Production
 (III) Nuclear Breeder Reactors – Selection, Testing, Large-Scale Deployment
 (IV) Coal Use Reassignments; Coal→ Nuclear Plant Modifications; Organic Chemicals Syntheses

Specifically, major research and development (R&D) in the next twenty years must include the following:[11]

8.1. INVESTIGATIONS OF SYNTHETIC FUEL (SYNFUEL) MANUFACTURE, STORAGE, DISTRIBUTION, AND SELECTION

(a) Synfuel production of hydrogen (H_2) from water (H_2O) + nuclear heat or electricity; by sulfur-iodine catalysis, electrolysis, and other economic methods.
(b) Synfuel production of ammonia (NH_3) and hydrazine (N_2H_4) from water (H_2O) + air (N_2) + nuclear heat or electricity. Also synfuel stabilization (e.g. hydration) of N_2H_4.
(c) Hydrocarbon and synfuel production from coal (C) + water (H_2O) + nuclear heat or electricity. Production of SASOL-process synpetrols (C_mH_n), butane (C_4H_{10}), acetylene (C_2H_4), etc.
(d) Alcohol synfuel production from agri-crops (e.g. corn) + sunshine + nuclear heat or electricity. Production of ethanol (C_2H_5OH), methanol (CH_3OH), etc.
(e) Bladder/adsorption storage materials for gaseous synfuels H_2, NH_3, CH_4, etc.
(f) Special high-pressure tanks for hydrogen (H_2) synfuel storage in automobiles and aircraft.
(g) Distribution of hydrogen and ammonia gas; synfuel filling stations and transfer techniques.

While pure hydrogen would be the preferred synfuel in automotive fuel-cell engine applications, the requirement for safe high-pressure tanks or adsorption bladders to reduce the storage volume of hydrogen, makes this gaseous fuel somewhat problematic at present. In addition, the countrywide distribution of hydrogen fuel poses problems. Unless storage and distribution difficulties for hydrogen can be solved in the next decade, ammonia synthesized from air and water looks more attractive as a portable synfuel for automobiles, since it can be safely confined as a liquid under a modest pressure of 20 atm in a fuel tank. Bio-alcohol, which is liquid and more similar to petrol, would be preferred for internal combustion engines (ICEs) if it were not for the extremely large areas of farm land needed for continuous biomass cultivation to satisfy the demands of our vast transportation fleets. Together with the availibilty of adequate amounts of nuclear elecricity or heat, bio-alcohol production may still be an affordable solution for oil replacement in countries with large areas of thinly populated arable lands, such as Brazil and the USA. Brazil has apparently chosen this path to

---

[11] Under the dedicated leadership and foresight of William Magwood, the US Department of Energy recently launched several initiatives for R&D on advanced (breeder) reactors and synfuel production (hydrogen, etc.) from nuclear heat and electricity. While encouraging, the US House and Senate need to expand such funding considerably and (re)start advanced breeder reactor development as a national priority.

overcome the pending out-of-oil crisis by expanding both nuclear power generation and bio-alcohol production. As discussed earlier, emissions of carbon-dioxide by burning bio-alcohols is exactly balanced by the intake of carbon-dioxide during bio-fuel plant growth, so there is no net atmospheric increase of this greenhouse gas. Ammonia used as a fuel in ICEs is technically feasible, but requires NOx suppression. This problem appears solvable with modern catalytic converters however. Ammonia (NH_3) can also be used in fuel-cell engines (FCEs) where it must first be catalytically decomposed to hydrogen (H_2) and nitrogen (N_2) on the FCE proton exchange membrane (PEM) to allow passage of pure hydrogen through it. Liquid hydrazine (N_2H_4) can similarly be used in a fuel-cell after catalytic decompositon into hydrogen and nitrogen. However hydrazine is less stable than ammonia and less desirable for public use, unless a good stabilizing agent can be found. Catalytic decomposition of ammonia or hydrazine by ruthenium alloys to provide hydrogen on fuel-cell PEMs has been demonstrated. There is no generation of NOx in low-temperature-operated FCEs.

Fuels produced under item 8.1(c) would irreversibly add carbon-dioxide to our biosphere from engine exhausts. This overburdens the atmosphere and according to predictions causes excessive global warming. Nevertheless, studies under item 8.1(c) are useful for developing new pathways to make hydrocarbons for plastics and other organic materials. SASOL synfuels may also assist in overcoming future oil shortages by bridging the critical period between present and future synfuel/engine technologies.

8.2. DEVELOPMENT OF ADVANCED AUTO/AIRCRAFT ENGINES AND OTHER ENERGY SUPPLYING DEVICES

(a) New ICEs burning H_2, NH_3, N_2H_4, C_2H_5OH synfuels with pure N_2 and H_2O or CO_2 exhausts (no No_x).
(b) Fuel-cell engines, consuming hydrogen (H_2), ammonia (NH_3), methane (CH_4), or hydrazine (N_2H_4).
(c) Improved low-weight portable electric storage batteries for vehicular propulsion.
(d) Improved low-weight portable flywheel devices (mechanical batteries) for vehicular propulsion.

Electrical and mechanical batteries (items 8.2(c) and 8.2(d)) look quite attractive at first glance. But nature does not allow electrochemical energy storage in a given mass and volume to exceed that obtainable from the combustion of a portable synfuel or petrol of the same mass and storage volume. To achieve a given travel range for an automobile (e.g. 600 km), one needs ten to a hundred times more mass if powered by a freshly charged battery than if propelled by a portable synfuel consuming engine. This restricts battery devices 8.2(c) and 8.2(d) to short-range (inter-city) applications in mass transportation. Still, it is useful to advance 8.2(c) and 8.2(d) for automotive applications where batteries can be recharged repeatedly every hundred kilometers with wall-plug electricity.

8.3. NUCLEAR BREEDER REACTOR DEVELOPMENT, TESTING, AND DEPLOYMENT

(a) Uranium → plutonium breeders
(b) Thorium → uranium breeders

As discussed in Subchapter 1.2, the biggest challenge will be to have a thousand breeder reactors deployed in the USA (and nine thousand world-wide), by the end of the next three decades. Within the next five to ten years, one or two basic breeder reactor designs must be selected for mass deployment. They must be built and fielded on a scale comparable to the WW-II mass-production of transport ships in the USA, but with stringent safety and durability features. After selection of an optimum breeder model, a comprehensive fast-track construction and deployment program must be formulated and integrated with the introduction of new synfuel engines and synfuel mass-production technologies. It will be necessary to determine early on whether nuclear heat or electricity (or a mix) is preferred for manufacturing synfuels. This can be investigated in pilot plant operations with heat from non-nuclear sources. Heat-generating breeder reactors could probably be built more quickly than electricity generating units which require large steam turbines and sophisticated electrical components. Both uranium-to-plutonium and thorium-to-uranium-233 breeder reactors should be explored (Section 5.1.2). A strong leader must be appointed to conduct this war-like effort. Brief 25 illustrates the US nuclear reactor expansion program that is necessary to stave off a collapse of our civilization by 2030.

Breeder reactors can consume depleted uranium which has been accumulating over the last twenty years at uranium enrichment plants that presently serve burners. Thus initially there is little need to increase uranium mining activities until these reserves are depleted. This is a bonus not available to the coal-burning option (see Subchapter 1.2) which must vastly expand mining operations. Also the final nuclear waste produced by breeders is much less than that created by burner reactors, so underground nuclear waste storage facilities can be utilized for longer periods of time (hundreds of years) before they are filled.

8.4. COAL USAGE REASSIGNMENT PROGRAMS

(a) Conversion of coal power plants to nuclear power plants with retention of existing steam turbines.
(b) Synthesis of organic compounds from coal, for plastics and other organics-dependent industries.

Coal-burning electric power plants may be converted into uranium-burning plants since steam generators are similar and usually housed in separate parts of the plant. Because many coal-burning power plants already exist, it would be worthwhile to explore the cost and feasibility of such a conversion when it is deemed essential to limit globe-warming carbon-dioxide emissions.

Organic chemicals presently extracted from crude oil, must be synthesized from coal as main feed material when oil is no longer available. Programs to provide

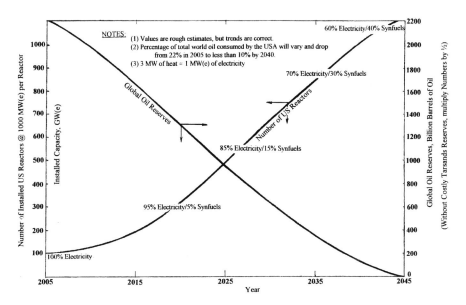

BRIEF 25. REQUIRED GROWTH OF U.S. NUCLEAR POWER PLANTS TO BALANCE GLOBAL OIL DEPLETION

new pathways for providing such organics must be ramped up and be ready to go into mass production by the year 2030. Some coal-burning electric power plants may be converted into heat-providing sources for the manufacture and synthesis of coal-derived organics until they can be replaced by uranium-breeding reactors.

8.5. ANTICIPATIONS

All R&D programs listed under Subchapters 8.1 through 8.4 should receive continuous government funding for at least the next three decades to be effective. Government labs, universities, and industries must all be engaged, and students must be trained to build and maintain a solid knowledge base. In addition, educational programs must be initiated to increase public understanding and acceptance of nuclear energy (Chapters 5–7).

It is important to re-emphasize that we are *not* considering a mere expansion or improvement in clean electric power generation. We are faced with the necessity to have petrol substitution fielded by 2030, and to supply *all* of man's future energy needs by 2050, that is electricity + portable fuels, without the use of oil, natgas, or coal. This leaves us with only uranium as a non-renewable prime energy provider. Proposals to expand renewable energy sources (solar, wind, and biomass) can only be supplemental for urban and rural electric power. Domestic electricity represents only 35% of the total energy pie (see Brief 26).

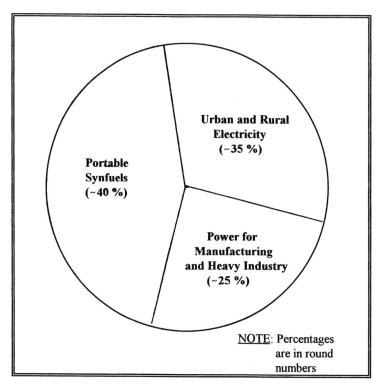

BRIEF 26. FUTURE GLOBAL ENERGY PIE

Denmark recently claimed to have reached its maximum capacity for windpower generation, which supplies 20% of its electric grid. However this is only 7% of its *total* energy consumption if one includes its use of petrol, diesel, and natural gas. Clearly when the latter energy sources are gone, it must find other means of replacing them to empower its transportation fleets and heavy industries. The only sources for additional prime energy are then coal or uranium whose production of electricity or heat allows the manufacture of synfuels, or the charging of portable energy units (storage batteries and flywheels) on a Joule for Joule basis. The laws of energy conservation and energy inter-conversions first enunciated by Sir James Prescott Joule and Nicolas Carnot are immutable: nothing comes for nothing. In summary, preferably by the year 2030 but not later than 2050, the entire energy pie shown in Brief 26 must be accomodated by nuclear fission (available for at least 1300 years), if one wants to avert an energy catastrophe and avoid global warming (i.e. no burning of coal).

Whether we can pass by the year 2030 unscathed without serious economic upheavals will depend on how forward-looking our future government leaders will be. They must distance themselves from the political influence of myopic extremists who insist that we slow or terminate nuclear power development and replace it

CONCLUSIONS, ACTION ITEMS, AND PREDICTIONS

NO MORE OIL !!??

entirely with solar cells and wind mills. Instead, government leaders must heed warnings of the nuclear energy engineering profession. Otherwise disastrous energy shortages will surely develop resulting in similar finger-pointing as with the 9/11 terrorist attacks on New York's World Trade Center, or the dam breaches around New Orleans from hurricane Katrina on August 29, 2005. Many early warnings were ignored. History shows time and time again that people and government bureaucracies tend to keep their heads in the sand till jolted. Unless remedied on time, oil depletion will produce a downward spiral of rapidly escalating shortages of food and goods, collapse of the economy, serious impoverization of the majority of people, and an increase of war-provoking world tensions. Desert cities like Las Vegas, Nevada and Phoenix, Arizona which are totally dependent on a cheap-oil economy, will become ghost-towns [Ref 53]. Construction of a nuclear power plant requires eight years of planning, designing, testing, and safety analyses, before it produces the first kilowatt. We cannot afford the luxury of meditation until catastrophy hits. But if history repeats itself, we probably will suffer great losses first before action is taken.

The sad part is that an escalating energy crisis is avoidable. Wise governments should start immediately investing in the synfuel research and development programs listed above and concurrently commence the design, development, and construction of avanced breeder reactors. As a bonus, it brings many new high-tech jobs and reduces unemployment, much like president Franklin Roosevelt accomplished with his national hydro-electric and road-work programs during the great depression of the 1930's. Environmentalists should join nuclear engineers to demand expansion of abundant non-air-polluting nuclear energy. Benjamin Franklin said at the signing of the Declaration of Independence in 1776: 'We must all hang together, or assuredly we shall all hang separately'.

ANNOTATED BIBLIOGRAPHY

BOOKS ON (NUCLEAR) ENERGY

[1] David Goodstein, *Out of Gas: The End of the Age of Oil*, W.W. Norton & Co, Inc, New York, London, (2004); ISBN 0-393-05857-3.
[2] Kenneth S. Deffeyes, *Hubbert's Peak; The Impending World Oil Shortage*, Princeton University Press, (2001); ISBN 0-691-116253.
[3] P. Roberts, *The End of Oil; On the Edge of a Perilous New World*, Houghton Mifflin Co, (2004); ISBN 0-618-23977-4.
[4] Heaberlin, S.W., *A Case for Nuclear-Generated Electricity*, Batelle Press, (2003); ISBN 1-57477-136-1.
[5] Joonhong Ahn and William E. Kastenberg, Editors, *Energy Future and the Nuclear Fuel Cycle in the Asia/Pacific Region*, Proceedings of Nuclear Energy Section of 19th Annual ILP Conf., U of Calif., Berkeley, College of Engineering and College of Chemistry (1997).
[6] Alan E. Waltar, *America the Powerless; Facing our Nuclear Energy Dilemma*, Cogito Books (Medical Physics Publishing), Madison WI, (1995); ISBN 0-99483-58-8.
[7] Richard Rhodes, *Nuclear Renewal; Common Sense about Energy*, Whittle Books/Viking Penguin, New York, (1993); ISBN 0-670-85207-4.
[8] Dixie Lee Ray, *Environmental Overkill; Whatever happened to Common Sense*, Harper Perennial, (1993); ISBN 0-06-09758-9.
[9] Edward Teller, *Energy from Heaven and Earth*, W.H. Freeman & Co, San Francisco, Oxford, (1979); ISBN 0-7167-1064-1.
[10] 10. C.T. Ellis, *A Giant Step - History of Rural Electrification in the USA*, Random House N.Y., (1966); Lib. Cong. Cat. No. 66-11991.
[11] Richard Rhodes, *The Making of the Atomic Bomb*, Simon & Schuster (1986); ISBN 0-671-44133-7.
[12] J.M. Blair, *The Control of Oil*, Random House N.Y., (1976); ISBN 0-394-72532-8.
[13] J. Gueron, J.A. Lane, I.R. Maxwell, J.R. Menke, editors, *The Economics of Nuclear Power*, McGraw-Hill, (1957).
[14] M. Cheney, *Tesla, Man out of Time*, Dell Publ. N.Y., (1983); ISBN 0-440-39077-X.
[15] *2002 World Almanac and Book of Facts*, Primedia, (2001); ISBN 0-88687-848-10.

BOOKS, REPORTS, AND ARTICLES ON FUEL-CELLS AND SYNFUELS

[16] J.O'M. Bockris and S. Srinivasan, *Fuel Cells: Their Electrochemistry*, McGraw-Hill (1969); Lib. Cong. 68-1475306345.
[17] H.A. Liebhafsky and E.J. Cairns, *Fuel Cells and Fuel Batteries*, John Wiley, (1968); Lib Cong. 68-22892; ISBN 47153420X.
[18] S. M. Hail, 'Materials for Fuel Cells', *Materials Today (Elsevier)*, p 24, March 2003.
[19] G. Marsh, 'Membranes fit for a Revolution', *Materials Today (Elsevier)*, p 38, March 2003.
[20] N. Sammes, 'Hot-Bed of Fuel Cell R&D', *Materials Today (Elsevier)*, p 44, March 2003.
[21] F.E. Pinkerton and B.G. Wicke, 'Bottling the Hydrogen Genie', *The Industrial Physicist*, p 20, February/March 2004.

[22] E.R. Riegel, *Industrial Chemistry*, 5th edition, Reinhold Pub Co, (1949).
[23] R.N. Shreve, *Chemical Process Industries*, 3rd ed, McGraw-Hill, (1967), Lib Cong No 68-20721.
[24] W. Braker, A.L. Mossman, *Matheson Gas Data Book*, Sixth Ed (1980), Lib Cong No 80-83159.
[25] T.J.W. Van Thoor, *Chemical Technology: An Encyclopedic Treatment Vol I*, Barnes and Noble Inc, (1968); Lib Cong No 68-31037.

BOOKS, REPORTS, AND ARTICLES ON NUCLEAR REACTORS, FUELS, AND RADIATION

[26] A. Tammaro, Editor, *The Shippingport Pressurized Water Reactor*, Addison-Wesley, (1958); Lib Cong No 58-12595.
[27] A.W. Kramer, Editor, *Boiling Water Reactors*, Addison-Wesley, (1958); Lib Cong No 58-12603.
[28] C. Starr and R.W. Dickinson, *Sodium Graphite Reactors*, Addison-Wesley, (1958); Lib Cong No 58-12598.
[29] J. Leclercq, *The Nuclear Age*, Publ. by Le Chêne, distributed by Hachette (France); Sodel 1986.
[30] M. Benedict, T. Pigford, H. Levi; *Nuclear Chemical Engineering*, McGraw-Hill, (1981); ISBN 0-07-004531-3.
[31] S. Glasstone, *Principles of Nuclear Reactor Engineering*, Van Nostrand (1955); Lib Cong No 55-8832.
[32] A.E. Waltar and A.B. Reynolds, *Fast Breeder Reactors*, Pergamon Press (1981); ISBN 0-08-0259820.
[33] S. Levine, Editor, *Reactor Safety Study*, WASH-1400 Rasmussen Report, US-ERDA, (Oct 1975).
[34] S. Rippon, 'Chernobyl: The Soviet Report', *Nuclear News*, pp 59-66, Oct 1986; Monthly Publication of American Nuclear Society.
[35] T.D. Lucky, *Hormesis with Ionizing Radiation*, CRC Press Inc. (1980).
[36] *Nucleonics*, McGraw-Hill, N.Y., Monthly Journal, 1957-1967.
[37] ANS (American Nuclear Society), *Nuclear News*, Monthly Issues 1961-2004.
[38] AIP (Am. Inst. of Physics), *Physics Today*, Selected Issues on Nuclear Topics, 1970-2004.
[39] ACS (Am. Chem. Society), *Chemical & Engineering News*, Selected Issues on Nuclear Topics, 1970-2004.

REPORTS AND ARTICLES ON COAL, OIL, AND NATURAL GAS

[40] J. Johnson, 'Getting to Clean Coal', *Chemical & Engineering News*, p 20 ff, Feb 23, 2004.
[41] T. Appenzeller, 'The End of Cheap Oil', *National Geographic*, p 84 ff, June 2004.
[42] C. Forest, M. Webster, J Reilly, 'Narrowing Uncertainty in Global Climate Change', *The Industrial Physicist*, p 20-23, August/September, 2004.

REPORTS AND ARTICLES ON WIND, SOLAR, AND GEOTHERMAL ENERGY

[43] J. Winters, 'Alternative Power Systems', American Society of Mechanical Engineers (ASME), *Mechanical Engineering*, Vol. 125, 1, pp 36-39, Jan 2003.
[44] J. Johnson, 'Blowing Green', *Chemical & Engineering News*, p 27-30, Feb 24, 2003.
[45] L. Kazmerski, T. Surek, 'Power from the Sun', Society of Photo Instrumentation Engineers (SPIE),*OEM Magazine*, pp 24-27, April 2003.
[46] P.B.Weisz, 'Basic Choices and Constraints on Long-Term Energy Supplies', American Physical Society (APS), *Physics Today*, pp 47-52, July 2004.
[47] Prospectus of Geodynamics, Ltd, 'Power from the Earth', ABN 55 095 006 090, August 2002, PO Box 2046, Milton, QLD 4064, Australia.

REPORTS AND ARTICLES ON BIO-MASS ENERGY

[48] A.P.C. Faaij, 'Energy from Biomass', PhD Thesis Univ. of Utrecht, Sep 29, 1997.
[49] C. Eerkens, 'An Holistic Approach to Evaluate the Potential Productivity of Unconventional Crops', pp 104-114, *Alternative Uses for Agricultural Surpluses*, edited by W.F. Raymond and P. Larvor; Elsevier Applied Science, 1986.

BOOKS ON RISK ANALYSIS

[50] P.C. Stern and H.V. Fineberg, editors; *Understanding Risk, Informing Decisions in a Democratic Society*; Nat. Academic Press (1996); ISBN 0-309-08956-5.
[51] D. Daniel and J. Applegate, editors; *Risk and Decisions about Disposition of Transuranic and high-level Radioactive Waste*; Nat. Academic Press (2005); ISBN 0-309-09549-2.
[52] C. Vignial, A. Monti, C. Dehouck, and E. Smiley, editors; *Terrorism Risk Insuarance in OECD Countries*; OECD Publishing (2005); ISBN 92-64-00872-1.

RECENT (POST-SCRIPT) BOOKS

[53] J.H. Kunstler, *The Long Emergency; Surviving the Converging Catastrophies of the Twenty-First Century*, Atlantic Monthly Press (2005), ISBN 0-87113-888-3.
[54] A.E. Waltar, *Radiation and Modern Life; Fulfilling Marie Curie's Dream*, Prometheus Books (2004); ISBN 1-59102-250-9.

ACKNOWLEDGEMENTS

I am grateful to my wife Els and my children Laura, Jelmer, Mieke, and Boukje, who stood by me during some difficult years in my professional and academic career. My daughter Mieke, who is an accomplished writer with a degree in English literature, edited and improved the readability of some sections in the book. It was also with great pleasure to work with Nathalie Jacobs and Anneke Pot of SpringerNL, who were very helpful and professional in helping to publish this book.

To Dick R. Griot, my former business partner and long-time friend, I am indebted for helping me stay the course in the promising and fascinating field of nuclear energy which I had chosen to devote myself to. Further I wish to express my gratitude to Jay F. Kunze, Dean of Engineering at Idaho State University, who encouraged me and kept me going. Finally I wish to express my appreciation for valuable discussions with colleagues Bill H. Miller, Sudarshan Loyalka, Mark Prelas, Tushar Gosh, and Bob Tompson at the Nuclear Science and Engineering Institute of the University of Missouri, Columbia. Others who saw rough drafts of my book and made specific suggestions to improve the book were: professors Tom Marrero and Allen Hahn of the University of Missouri; Dr. Bob Gordon, retired president of SCI, Inc, and a former nuclear engineering class-mate; Steven Barnholtz a former science teacher from Pacific, Missouri; Dr. Bertus de Graaf, vice-president of Geodynamics in Australia; Dr. Piet Borst, professor of molecular biology, Netherlands Cancer Institute, Amsterdam; and Ir C. Eerkens, former agricultural consultant to the EU Agriculture Mission in Brussels, and staff engineer at Unilever, Rotterdam, and UNDP, NewYork.

SHORT BIOGRAPHY OF THE AUTHOR

Jeff W. Eerkens is an adjunct research professor at the University of Missouri in Columbia. He has a doctorate in Engineering Science (1960) and a masters degree in Nuclear Engineering (1957) from the University of California at Berkeley, and is a registered nuclear engineer in the State of California. His doctoral dissertation was a study of the chemical effects in fluids produced by fission fragments of uranium. His Ph.D. work included graduate studies in biochemistry and the origin of life.

Dr. Eerkens has had extensive "hands-on" experience with nuclear reactors, isotope separation systems, and various aspects of the nuclear fuel cycle. Prior to research on laser separation of medical isotopes in Missouri, he spent thirty years in California working in staff positions for several aerospace companies. He participated there in the design and testing of nuclear and solar energy systems for space applications. This included the preparation of reactor hazards analysis reports and the execution of reactor experiments.

In 1957, as a student at UC-Berkeley's Engineering Field Station, Eerkens measured molecular (isotope) separations in annular shock regions of free jets, a variant of a nozzle isotope enrichment method developed earlier by Becker in Germany. In 1959/1960 he carried out critical mass assemblies of two research reactors, one at the University of Oklahoma, and one at Rice University in Texas. In 1964/1965 Eerkens designed and operated the world's first nuclear-pumped gas laser by exciting lasable noble gases directly with uranium fission fragments in a laser tube placed inside a pulsed TRIGA reactor at the Norhtrop Space Laboratories. In 1972/1973 he initiated experiments to study isotope enrichment of gaseous uranium hexafluoride using a molecular laser. This ultimately led to a new laser isotope separation technique called CRISLA (Condensation Repression Isotope Separation by Laser Activation) which employs a supersonic free jet and laser-activated condensation repression of selected isotopomers.

Dr. Eerkens holds the first patent on a gamma-ray laser or 'graser', and is a co-patent-holder of the first green Helium-Neon laser sold commercially. In addition to experience in the nuclear and laser fields, he performed studies on rocket plume infrared radiation signatures for the US Air Force, and participated in designing infrared optical systems and space power units for aerospace satellites. He also designed and built high-power lasers, and investigated radiation hardening of optical materials.

Dr. Eerkens is of Dutch descent and was born in Indonesia (formerly the Dutch East Indies), where his father was a physician. As a child during World War II, he spent three years in a Japanese concentration camp on Java. The atomic bomb that ended WW-II saved his life (and of half a million others in Asia). It later spurred his interest in nuclear physics and engineering. After enrolling at the University of California in Berkeley as a foreign student in 1950, he obtained his Ph.D. degree in 1960 and became a US citizen in 1961. He lived in California till 1994, in Missouri till 2005, and returned to California after that. He is married and the father of four children.

ABBREVIATIONS

AC	Alternating Current (Electricity)
AEC	Atomic Energy Commission ($\to$ DOE)
AVLIS	Atomic Vapor Laser Isotope Separation
BTU	British Thermal Unit
BWR	Boiling Water Reactor
CANDU	CANadian Deuterium Uranium (Heavy Water Moderated and Cooled Reactor)
DC	Direct Current (Electricity)
DIF	Diffusion, Process for Uranium Enrichment
DOE	Department of Energy (USA)
DOT	Department of Transportation (USA)
DNA	Desoxyribo Nucleic Acid (Biomolecule)
FP	Fission Product
ECCS	Emergency Core Cooling System
GDP	Gross Domestic Product
GE	General Electric (Corporation)
GM	Geiger Müller (Radiation Counter)
HP	Horse Power (Unit for Energy Consumption Rate = 746 Watts)
HP	Health Phycists (monitors radiation levels around nuclear reactors)
IAEA	International Atomic Energy Agency
IC	Internal Combustion
ICE	Internal Combustion Engine
IN(E)L	Idaho National (Engineering) Laboratory
ITER	International Thermonuclear Experimental Reactor
KWI	Kaiser Wilhelm Institute (Discovery of Uranium Fission)
LANL	Los Alamos National Laboratory
LIS	Laser Isotope Separation
LLNL	Lawrence Livermore National Laboratory
MLIS	Molecular Laser Isotope Separation
MCA	Maximum Credible Accident
MPD	Maximum Permissable Dose
MPE	Maxiumum Permissable Exposure
NASA	National Aeronautics and Space Administration (USA)
NEI	Nuclear Energy Institute
NPT	Non Proliferation Treaty
NRC	Nuclear Regulatory Commission (USA)
OPEC	Organization of Petroleum Exporting Countries
PEM	Proton Exchange Membrane
PF	Pons-Fleischman
PWR	Pressurized Water Reactor
RNA	Ribo Nucleic Acid (Biomolecule)
RBE	Relative Biological Effectiveness

SASOL	South Africa Synthetic Oil Ltd
TBP	TriButyl Phosphate
UCF	UltraCentriFuge, for Uranium Enrichment
URENCO	Uranium Enrichment Company
USEC	United States Enrichment Corporation

CHEMICAL SYMBOLS OF SELECTED ELEMENTS AND ISOTOPES

B	Boron (Z = 5); B-10 = Boron (M = 10, Z = 5), B-11 = Boron (M = 11, Z = 5)
C	Carbon (Z = 6); C-12 = Carbon (M = 12, Z = 6), C-13 = Carbon (M = 13, Z = 6)
H	Hydrogen (Z = 1); H-1 (M = 1, Z = 1)
H-2	D = Deuterium (M = 2, Z = 1)
H-3	T = Tritium (M = 3, Z = 1)
He	Helium (Z = 2); He-4 = Helium (M = 4, Z = 2), He-3 = Helium (M = 3, Z = 2)
K	Potassium or Kalium (Z = 19); K-39 (M = 39, Z = 19), K-40 (M = 40, Z = 19), K-41 (M = 41, Z = 19)
Li	Lithium (Z = 3); Li-6 = Lithium (M = 6, Z = 3), Li-7 = Lithium (M = 7, Z = 3)
N-14	Nitrogen (M = 14, Z = 7), N-15 (M = 15, Z = 7)
O-16	Oxygen (M = 16, Z = 8); 0-18 (M = 18, Z = 8)
Th-232	Thorium (M = 232, Z = 90)
U-233	Uranium (M = 233, Z = 92); U-235 = Uranium (M = 235, Z = 92)
U-238	Uranium (M = 238, Z = 92)
Np-239	Neptunium (M = 239, Z = 93)
Pa-233	Protactinium (M = 233, Z = 91)
Pu-239	Plutonium (M = 239, Z = 94)
I-127	Iodine (M = 127, Z = 53); I-126 (M = 126, Z = 53)
Cs-133	Cesium (M = 133, Z = 55); Cs-137 (M = 137, Z = 55)
Sr-88	Strontium (M = 88, Z = 38); Sr-90 = Strontium (M = 90, Z = 38)
Hg-200	Mercury (M = 200, Z = 80)

NOTE:

M = Atomic Mass Number = Number of protons + number of neutrons in nucleus of atom
Z = Atomic Charge Number = Number of protons in nucleus = Number of electrons in atom

ATOMIC/MOLECULAR ENERGIES:

1 eV (electron-Volt) = 1.6021×10^{-19} Joule per atom or molecule = $9.65 \times 10^4/M$, Joule per gram
1 MeV = 10^6 eV = $1.6021 \times 10-13$ Joule per atom or molecule = $9.65 \times 10^{10}/M$, Joule per gram
1 gram of an element with mass number M contains N = $6.0247 \times 10^{23}/M$ atoms or molecules

INTERNATIONAL MKS UNITS AND PREFIXES

INTERNATIONAL MKS UNITS

m	meter, unit of length
kg	kilogram, unit of mass
s	second, unit of time
J	Joule, unit for energy
W	Watt (= J/s), unit for rate of energy consumed/produced per second
Gr	Gray, unit for radiation dose (1 Joule deposited radiation energy per kilogram biomatter)
Sv	Sievert, unit for man-effective radiation dose (1 Gray × Biological Effectiveness Factor)

PREFIXES

a	atto $= 10^{-18}$, quintillionth part
f	femto $= 10^{-15}$, quadrillionth part
p	pico $= 10^{-12}$, trillionth part
n	nano $= 10^{-9}$, billionth part
μ	micro $= 10^{-6}$, millionth part
m	milli $= 10^{-3}$, thousandth part
c	centi $= 10^{-2}$, hundredth part
k	kilo $= 10^{3}$, thousand-fold
M	mega $= 10^{6}$, million-fold
G	giga $= 10^{9}$, billion-fold
T	tera $= 10^{12}$, trillion-fold
P	peta $= 10^{15}$, quadrillion-fold
E	eta $= 10^{18}$, quintillion-fold

INDEX

AC (Alternating Current) Electricity, 38, 42
accidents, auto crashes, 5, 6
accidents, nuclear
   Chernobyl, Russia, 7, 26–28, 114–116
   malfunctions and mishaps, 109
   serious, 112–116
   submarines, 110
   Tokaimura, Japan, 111
   TMI (Three-Mile-Island), USA, 7, 27, 112
amides, mixed; for hydrogen storage, 66
air pollution,
   by coal, 25, 88
   by automobiles (petrol), 8, 53, 63, 66
air + water + electricity fuel
   synthesis, 56, 132
airplane flight, radiation exposure, 106
airplane crash, on reactor, 7, 73, 115
aircraft (aviation) propulsion with synfuels, 53
Alamogordo, New Mexico (first atomic
   bomb test), 45, 46
alcohol, as synfuel, 8, 13, 16, 39, 40, 52,
   132, 133
alpha particle, radiation/emission, 97
ammonia (NH_3), as synfuel, 8, 9, 11, 24, 34, 52,
   53, 56, 63, 66, 132
Anglo-Persian/Anglo-Iranian Oil Company, 43
anti-nuclear (sentiments, activities), 2, 3, 4, 8,
   19–29, 38, 136, 137
anti-radiation tablets, pills, 29, 114, 115
AREVA, French Consortium handling uranium
   enrichment and fuel cycle, 86, 119
atoms, constitution of, 44, 101
auto (automobile), fuels and engines for, 51ff
atmosphere, pollution of, 25, 53, 66, 88
Atkinson, R. d'E (nuclear fusion pioneer), 93
AVCO (company which explored AVLIS), 87
aviation, future fuels for, 53
AVLIS (Atomic Vapor Laser Isotope Separation)
   for uranium enrichment, 87

badges, for radiation dose monitoring, 107
battery, electric storage for auto propulsion, 51,
   52, 57, 58
benzin (German for petrol), 1
beryllium, neutron moderator material, 73

beta, radiation/emission, 72, 77, 79, 97–104, 108
beta, effects on bio-organisms, 100–102
Bethe, Hans (nuclear fusion pioneer scientist), 93
bio-effects, by radiation, 99–103
biofuel/biomass, energy from, 5, 12, 13, 24,
   38–40, 48–49
bladder, for hydrogen gas storage, 23, 55, 66, 132
Bohr, Niels (pioneer atomic and nuclear
   physicist), 44, 45
bomb, attack on reactor, 125
bombs, nuclear, 19–20, 79–81, 123–125
bone marrow, radiation effects, 106–107
Brazil, energy program of, 132
breeder reactors, 5, 10, 11, 17, 22, 35,
   77–79, 134
breeding, to make fisionable isotopes, 22, 77–79
   uranium-238 → plutonium-239, 79
   thorium-232 → uranium-233, 79
British Petroleum (BP) Oil Company, 43
BTU = British Thermal Unit, Energy Unit in
   English System, 33, 34
burner reactors, "burners", 10, 17, 22, 35, 72, 78
burn, by gammas, 100
burning, (air-polluting and wasteful) of coal, oil,
   natgas 25, 35–36, 61–62
butane, in natural gas, 35, 52
BWR (Boiling Water Reactor), 72

Cal (= 1000 calories = 4,186 J), unit of energy
   used in dietetics, 32
cancer
   cause of, 102
   from fall-out radiation, 28
   from Chernobyl, 28
CANDU (CANadian Deuterium Uranium
   Reactor), 73
canisters, for radiowaste, 82–84, 88
carbon dioxide,
   exhaust and global-warming by, 14, 18, 24, 25,
      40, 53, 55, 56, 58, 63, 66, 133, 134
   as reactor coolant, 76
carcinogens, 102–103
Carnot, Nicolas L. S. (formulator of energy
   conversion law), 33, 136

cars, propulsion of, 33, 51, 61–67
caskets, for spent fuel transport, 25, 82–84, 88
casualties,
    from Chernobyl, 28, 114
    from Three Mile Island (TMI), 28, 114
catastrophe, no-oil, 1–4
cesium-137, fission product radioisotope, 99
chain reaction, by neutron-induced fission, 70–72, 80
Chandrasekhar, S. (pioneer astrophysicist), 93
chemicals, toxic, 25, 102
Chernobyl, reactor acccident, 7, 27–29, 71, 113–116
Chevron (Standard Oil of California) oil company, 43
Chrysler (electric and fuel-cell car maker), 61, 66
clad, cladding, in nuclear fuel element, 73, 75, 82
clathrates, for hydrogen gas storage, 66
clean-up, of radioactive fallout, 29, 124
cloud, radioactive dust, 29, 98, 114
coal, 5, 14, 15, 16, 22, 24, 25
    consumption of, 35
    depletion time, 14, 35–36
    earth reserves, 36
coal + water + electricity fuel synthesis, 55, 56, 132
coal, pollutants from burning, 14, 25, 69, 88
Cockroft-Walton particle accelerators, 93
coffee drinking, 28
Cohen, Bernard (engineering professor, nuclear hazards analysis specialist), 91
cold fusion, 94, 95
combustion engine, internal (IC, ICE), 51, 61–63
condensation, of steam, 74, 76
condenser, in steam turbine cycle, 74, 76
cooling tower, in power plants, 76,77
consumption, energy, 34–35, 136
containment vessel, around nuclear reactor, 25, 26, 29, 73, 112, 114
contamination, by radiation fallout, 26–29, 112–116
control, of nuclear weapons proliferation, 117–122, 125–129
control rod, in nuclear reactor, 74–76
conversion, methods of energy, 31–34, 37–38, 57–67, 69–71
core, of nuclear reactor, 73–77
cosmic radiation, radiation exposure from, 105, 106
cost, of electricity, 13–16, 20, 40–41
Coster, Dirk (Dutch physicist who helped Lise Meitner escape from Nazis), 44
crisis, out-of-oil, 2–5, 137, 138

critical, criticality of fissionable mass, 45, 70, 71, 74–76, 80, 81, 112, 113, 122–124
Curie, Marie (world-famous chemist, discoverer of radioactive elements), 82, 111
cycle, nuclear fuel, 81–92
cycle time, of neutron, 71

Daimler, Gottfried (first to put combustion engine in automobiles), 42, 61
D'Arcy, William (pioneer oil field developer), 43
dark energy, 131
DC (Direct Current) Electricity, 42
DC versus AC Electricity, 42
decay times,
    of plutonium-239 (24,000 y), 78, 90
    of radioisotopes, 27
    of uranium-238 (4.5×10^9 y), 27
delayed neutrons, 71, 74, 78
democracy, democratic world, 5, 16
Denmark, wind-power in, 121, 136
dental/medical, xray radiation dose, 106
depletion times, of earth energy resources, 35, 36
detectors, of nuclear radiation, 107
deuterium,
    as neutron moderator, 73, 78
    as fusion fuel, 93–95, 131
diesel, form of petrol fuel, 1, 52, 62
diffusion,
    gaseous in uranium enrichment, 86
    of radioisotope through soils, 91
dihydrogen oxide (H_2O), 19
disposal, of radioactive waste, 88–92
DNA (Desoxyribo Nucleic Acid) biomolecule, 99, 101–103
dosage (dose) of radiation, units and standards, 104–108
    lethal, 106
    maximum permissible, 105
dust, radioactive, 29, 98, 114, 124
duration, of earth energy resources, 2, 20, 35–37

earthquake, effect on reactor operation, 75
economy, collapse due to oil depletion, 2–5, 137
ECCS (Emergency Core Cooling System), 74, 112, 113, 125
Eddington, A.S. (pioneer nuclear astrophysicist), 93
Edison, Thomas (inventor of lightbulb and other electronic devices during 1880–1910), 42
Einstein, Albert (world famous physicist – gravitation theory), 45

INDEX 155

Eisenhower, Dwight D., (US president; initiated "atoms for peace" program), 118, 126
electricity,
  benefits to man, 38
  cost of, 13–16, 20, 40–41
electricity, AC versus DC, 42
electricity generation, 2–4, 69ff
  from coal, 5, 6, 20, 21, 69ff
  from geothermal, 20, 21
  from hydro, 20, 21
  from windmills, 5, 12–14, 20, 21
  from solar, 5 ,12–14, 20, 21
  from uranium fission, 5–12, 20, 21, 69ff
electric grid, 38, 42, 69, 70
electric storage battery, 52, 57–58
electrochemical fuel cell, 63–66
electrons, 38, 65, 97–104
elements, atomic, 44, 45, 52–56, 149
elements, reactor fuel, 73–75, 82, 84, 87
energy,
  equivalences, 33–34
  heat, 31–33
  kinetic, 31–33
  mechanical, 31–33
  potential, 31–33
energy,
  from electric battery, 52, 57–58
  from flywheel battery, 52, 58–61
  from fuel-cell, 63–67
  from internal combustion, 61–63
  from steam, 67
energy consumption (energy pie)
  in USA, 34–37
  global, 34–37, 136
energy consumption, 1–4, 34–37
  of electricity, 34–37, 136
  for transportation (portable fuels), 35, 36,136
  in homes, 36, 135, 136
  for industries, 35, 36, 136
energy conversion
  heat to electricity, 33, 34
  electricity to mechanical, 33, 34
  heat to mechanical, 33, 34
energy, dark, 131
energy, definition of, 31–34
energy crisis, from oil depletion, 2–5, 137, 138
energy depletion periods of earth resources, 35–37
  coal, 35–37
  natgas, 35–37
  oil, 35–37
  uranium and thorium, 35–37
  deuterium and lithium, 94

energy depletion, of geothermal heat, 21, 47, 48
energy liberation, in
  coal combustion, 36
  natgas combustion, 36
  petrol combustion, 36
  uranium fission, 36, 45
  deuterium fusion, 93, 94
energy resources, earth, 35–36, 135
enforcement, of NPT, 125–129
engineers, 1, 3, 4, 19, 91, 137, 138
enrichment, of uranium isotopes, 84–87
  by electromagnetics (calutron), 84, 85
  by gaseous diffusion, 84–87
  by laser excitation, 86, 87
  by ultracentrifuge, 86, 87
environmental pollution
  by coal burning, 14, 18, 25, 69, 88
  by burning petrol, natgas, 53, 63, 66, 133
  by uranium fission, 88–92
environmentalists, 13–15, 138
Esso/Exxon, 43
ethane, in natural gas, 35, 47, 52
ethanol (= alcohol), as synfuel, 8, 13, 16, 39, 40, 52, 132, 133
eV (electron-volt), atomic energy unit, 58, 64, 72, 149
excavation, for radiowaste storage, 83, 84, 88–92
excursion (prompt critical) of nuclear reactor power, 113
exposure, to radiation, 104–107

facts, about nuclear energy, 19–29
fables (fiction), about nuclear energy, 19–29
fallout radiation
  from nuclear bomb, 28, 29, 98
  from Chernobyl, 28, 29, 114
Fermi, Enrico (nuclear physicist who built first nuclear reactor in WW-II), 47, 70
fission, of uranium and plutonium by neutron induction, 72, 79
fission products, 72, 79, 98
Fleischman, M. and Pons, S.J. (cold-fusion experiments), 94, 95
flywheel, energy storage battery, 59–61
folklore, nuclear, 19
Ford, Henry (pioneer automobile maker), 42
Ford motor company, 57
fossil fuels, earth energy resource, 35–37, 39
FP1, FP2, fission products, 72, 79
Franklin, Benjamin (signer of US declaration of independence), 138
Frisch, Otto (Lise Meitner's nephew, chemist who first validated nuclear fission), 44, 45

fuels
  natural, 52
  synthetic portable, 52–57
fuel cell, 22–24, 51ff
  operating principle, 63–67
  automobile engine, 67
fuel cycle, nuclear (uranium), 83
fuel element, nuclear, 73–75, 82, 84, 87
  fabrication, 73–75
  in reactor core, 72, 82, 84, 87
Fulton, Robert (pioneer builder of first practical steam engine), 41
fuse, for reactor safety, 75
fusion, nuclear, 22, 92–95
  process, 92–95
  power plant, 94
furnace, high-temperature, in AVLIS process, 87

gamma radiation, 28, 73, 97–99
gamma dose, 104–107
gamma radiation, effect on biomatter, 99–104
Gamov, George (pioneer nuclear fusion scientist), 93
GJ (GigaJoule), 32
gas, natural (natgas), 35ff
  consumption, 35–37
  depletion times, 35–37
  reserves, 35–37
gasoline, (US name for petrol), 1
gas turbine, for electricity generation, 73
GDP (Gross Domestic Product), 21
GE (General Electric) company, 42
generation of electricity
  from bio-mass, 38–48, 48–49
  from coal, 69
  from geothermal energy, 21, 47–48
  from hydroelectric waterfalls, 21, 48
  from solar cells, from windmills/wind-turbines, 20, 21, 40, 48
  from uranium/plutonium, 70ff
geothermal energy, 21, 47–48
global warming, from carbon dioxide, 14, 18, 24, 25, 40, 53, 56, 58, 63, 66, 133, 134
GM (Geiger-Müller) radiation counter, 107
GM (General Motors) company, 57
Goffman, J.W. (anti-nuclear physician), 102
granite, radiation from, 106
Gray, radiation dose unit, 107
grid, electric (electricity distribution network), 38, 42, 69, 70
Grove, William (fuel cell pioneer), 63

Groves, Leslie (US Army general in charge of WW-II Manhattan Project), 11, 18, 84
Gulbenkian, C.S. (early Middle-East oil field speculator), 43
Gulf Oil Company, 43

Hanford, Washington, plutonium production plant, 45
Hahn, Otto (nuclear/chemical physicist, co-discoverer of uranium fission), 44, 70, 85
Harkins, W.D. (early nuclear fusion (astro)physicist), 93
hazards
  from radioactive wastes, 88–92
  from reactor accidents, 109–116
hazards analysis report, 76
health physicist (HP), resident in nuclear power plant, 108
heat, as form of energy, 31–33
helium gas, as reactor coolant, 73
helium ion (alpha particle), 93
Heisenberg, Werner (world renowned physicist – formulated quantum mechanics), 85
Hiroshima, atomic bomb, 46, 85
hormesis (Ref.35), 102
horsepower (HP), 32
Houtermans, F.G. (pioneer nuclear fusion (astro)phycist), 93
Hubbert (Shell engineer predicting fossil fuel extractions to peak by end of 20th century), 20
hydrazine (N_2H_4), as synfuel, 52–57, 63
hydrides, metal-, for hydrogen storage, 66
hydro-electric power plants, 21, 48
hydrogen (H_2), as synfuel, 22–24, 34, 52–57, 63–66
hydrogen fuel cell, 22–24, 34, 63–66
hydrogen, as neutron moderator, 72, 78

IAEA (International Atomic Energy Agency), 125–129
INL (Idaho National Laboratory), 71
inertial confinement nuclear fusion, 94
ingestion, of nuclear fallout, 29, 98, 114
inhalation, of nuclear fallout, 29, 98, 114
internal combustion engine (ICE), 51, 53, 61–63
iodine, as catalyst in hydrogen production, 55
iodine, radioactive in fission products, 29, 99
ionizing radiation, 99–103
ionization, 99–103
Iran, Iraq oil fields; Iraq Petroleum Company, 43

INDEX

isotopes, of atomic elements, 22, 44, 72, 77–79, 94, 99, 149
isotopes, of uranium, 22, 72, 73, 77–79, 149

Jaques, W.W. (fuel cell pioneer), 63
J (Joule), international energy unit, 32, 107
Joule, James Prescott (famous 19th century physicist), 31
jolt, from economic collapse, 99

K-40 (potassium-40) naturally radioactive element in blood, 27, 106
Kennedy, Robert (assassinated US presidential candidate), 38
KI (potassium iodide) tablet against radioiodine uptake, 29, 107, 114
knifecut, equivalent radiation dose, 102, 104, 105
kW (kilowatt), energy rate unit, 32–34
kWh (kilowatt-hour), energy unit, 32–34
KWI (Kaiser Wilhem Institute; discovery of uranium fission), 44

LANL (Los Alamos National Laboratory), US Nuclear Weapons Lab, 45
laser, in isotope separation (LIS), 86, 87
lesions, in biomolecules from ionizing radiation, 101–103
lethal radiation dose, 106, 107
lifetime
  of neutron, 71
  of radioisotopes, 72, 88–90
LIS (laser isotope separation) for uranium enrichment, 86, 87
lithium batteries, 57
Livermore, California, 87
LLL, LLNL (Lawrence Livermore (National) Laboratory), US Nuclear Weapons Lab, 87
Los Alamos, New Mexico, 45
Lucky, T.D. (pioneer investigator of hormesis; Ref. 35), 102

M (Atomic Mass Number) of Isotope or Element, 44, 45, 149
magnetic confinement nuclear fusion, 93
Manhattan project, 45
MCA (Maximum Credible Accident), 110–116
mechanical (flywheel) battery, 59–61
medical/dental xray radiation dose, 106
medicine, nuclear, 46, 71, 100, 108
Meitner, Lise (nuclear/chemical physicist, co-discoverer of uranium fission), 44–46, 70

meltdown, of reactor core, 7, 27, 110–116
mercury, polution from coal, 25, 88
methane
  as synfuel, 52, 56
  in natgas, 36, 52, 56
methanol, as synfuel, 52, 54
MeV (million electron-Volt), atomic energy unit, 45, 72, 93, 94, 97, 149
Middle East oil fields, 43
mine, for radiowaste storage, 88–92
MLIS, Molecular Laser Isotope Separation for uranium enrichment, 86, 87
Mobil Oil Company, 43
moderator for neutron thermalization, 72, 73
molecules, constitution of, 101
monitors, to measure radiation levels, 107
MPD (Maximum Permissable Dose), 105
MPE (Maximum Permissable Exposure), 105
mrem (man-equivalent rem) radiation dose unit, 104–105
mutations in bio-molecules, 99–103
MW (mega-watt), energy rate unit, 32–34

Nagasaki, Japan, atomic bomb, 46
NASA (National Aeronautics and Space Administration), fuel-cell use by, 63, 65
natgas (natural gas), 35–38
  depletion times, 35, 36
  consumption of, 35, 36
  earth reserves, 35, 36
  from seabeds, 35, 36
natural radiation sources, 106
naval nuclear reactors, 27, 110
negative reactivity, of nuclear reactor, 26, 27, 113, 116
neutron, 70ff
  birth in fission, 70, 71
  fast, 72
  thermal (slowed-down), 72–74
  lifetime of, 71
  multiplication in fission, 70ff
  moderation (thermalization) and reflection of 72ff
  poisoning (by neutron absorbers), 75, 76
Nevada, Yucca mountain repository, 10, 84, 92
Newcomen, Thomas (inventor of steam-driven suction pump), 41
Niagara Falls, first US hydroelectric power plant, 42
nickel battery, 57
Nissan motor company, 57
nitrogen gas as reactor coolant, 73
nitrogen oxides (NO_x), air pollutants, 24, 63, 133

nitroglycerine, 19
non-renewable energy, 45–49
NPT (Nuclear Non-Proliferation Treaty), 7, 125–129
nuclear
  energy, 2–4, 5, 6–12, 17–18, 19–29, 35–37, 72–79, 93, 94, 135ff
  engineers, 1, 3, 4, 19, 91, 137, 138
  fallout, 28, 29, 114
  fission, 72–79
  fuel cycle, 81–92
  fusion, 92–95
nuclear medicine, 46, 71, 100, 108
nuclear reactor
  power, 71–79
  research, 46, 71
nuclear power plants
  USA, 7, 11, 12, 17, 18, 21, 71
  Global, 11, 12, 17, 21, 71
nuclear weapons (bombs), 19, 79–81, 117–129
nucleus, composition of, 32, 44, 101

Oak Ridge, Tennessee, Uranium enrichment plant, 45, 85
oil, 1, 42, 43
  consumption rates, 42, 43
  depletion, depletion times, 1, 2, 15, 16, 30
  energy from, 35
  fields, 42, 43
  products from, 37
  reserves, tarsands, 34–38
Oppenheimer, Robert (nuclear physicist, head of WW-II Manhattan Project science staff), 45
Ostwald, W. (pioneer fuel-cell chemist), 63
Otto, N.A. (inventor of internal combustion engine "Otto cycle"), 61

Parsons, Charles (inventor of steam-driven turbo-generator), 42, 69
particle
  alpha, 28, 97ff
  beta, 28, 97ff
particle, radioactive dust, 29, 98, 114
PEM (Proton Exchange Membrane), 64–66, 133
Perrin, J. (early nuclear fusion (astro)physicist), 93
petrol (benzin, gasoline, diesel), definition of, 1
photon
  gamma, 28, 97ff
  xray, 97, 106
  ultraviolet (UV), 106
  visible, 97

plant, nuclear power, 69–79
plutonium (Pu-239), fissionable reactor fuel, 77–79
plutonium, theft of, 123–124
Pons, S.J. and Fleischman, M., (cold fusion experiments), 94–95
pollution
  from coal, 14, 18, 25, 88
  natgas and oil, 24, 53, 56, 58
  thermal, 76
  uranium fission, 88
poisoning, neutron (by neutron-absorber), 75, 76
Pottasium-40 (K-40), natural radiosotope in humans, 27, 105, 106
power, definition of, 32, 33
power, nuclear reactor, 69–79
proliferation, nuclear (NPT), 7, 125–129
propane, in natural gas, 35, 47, 52

rad, radiation dose unit, 104
radiation, 28, 29, 97–104, 106
  cosmic, 28, 106
  particles, alpha, 28, 29, 97–106
  particles, beta, 28, 29, 97–106
  photons, gamma, 28, 29, 97–106
radiation, biological effects of, 28, 29, 97–106
radiation detectors and monitors, 107
radiation dose
  airplane trip, 106
  background natural sources, 106
  cosmic (Colorado) rays, 106
  dental/medical xrays, 106
  exposure standards, 104–108
  potassium-40 (K-40) in blood, 27, 105, 106
radioactive waste (radio-waste), 26, 27, 82, 83, 88–92
radioisotopes (radioactive isotopes), 44, 46, 72
  for medical applications, 46, 71
  in fission products, 72, 99
  in fallout, 99
  naturally occurring, 99, 106
radiotherapy, 46, 71, 108
radium, radon, 105, 106
Rayburn, Sam (long-term speaker of US House of Representatives in mid-1900's), 4
RBE (Relative Biological Effectiveness), 104ff
reactivity, of nuclear reactor core, 25, 76, 81, 113
  negative temperature coefficient of, 25
  positive temperature coefficient of, 25, 113, 115
reactor control rods, 75, 76
reactor core, 72–75
reactor fuel elements, 72, 73

reactor refueling, 76
reactor shielding, 73, 74
reactors
   naval and submarine propulsion, 27, 110
   electric power generation, 7, 11, 12, 17, 18, 21, 71–79
   research, 46, 71
reactor types
   BWR, 72
   CANDU, 73
   gas-cooled, 73, 76
   PWR, 72
   RBMK, 115
recycling
   of plutonium, 77ff, 87, 88
   of water (H_2O), 72ff
reflector, neutron, 73
rem, radiation dose unit, 104
renewables, energy resources, 5, 12–14, 17, 20, 21, 38–41, 46
   biomass, 13, 48, 49, 39, 40, 46
   hydro, 21, 46
   solar and wind, 5, 12–14, 20, 21, 40, 41, 46, 48
repository, for nuclear waste, 83, 84, 88–92
reprocessing, of spent nuclear fuel, 87, 88
research reactors, for medical radioisotope production, 46, 47
reserves, of natural energy resources, 34–37
resources, of energy fuels, 34–37
Rhodes, Richard (Pullitzer-prize-winning writer), 44
Rickover, Hyman (Admiral and founder of US Nuclear Navy), 11, 18
Roosevelt, Franklin Delano (US president in WW-II), 45
Royal Dutch Shell, oil company, 43
runaway, reactor core power, 26, 111, 113–114

sabotage, of reactor operation, 74, 80, 125
safeguards
   against nuclear proliferation, 125–129
   against terrorists, 122–125
safety features, of nuclear reactor, 97–109
salt-bed mine, radiowaste disposal, 88, 89
SASOL (South Africa Synthetic Oil Ltd),
   coal → petrol conversion process, 54, 133
satellite, for NPT surveillance, 127
scram, reactor, 75
scratch (on skin), equivalent radiation dose, 28, 29, 104

sea-bed, natgas (methane) deposits, 35, 36, 37
sea-bed, radiowaste disposal, 88–89
security systems
   for NPT, 125–129
   prevention of nuclear fuel theft, 122–125
seeds, radioactive (in nuclear medicine), 71, 100, 108
shale/tarsands oil, 36, 37
Shell Oil Company, 43
shielding, reactor, of gammas, 73–74
shim control rods, of nuclear reactor, 76
Siegbahn, Karl (well-known Swedish nuclear physicist), 44
Sievert, radiation doese unit, 107
skin, penetration by particle radiation, 100
solar
   energy, 39
   radiation (UV), 106
South Africa (SASOL plant), 54, 133
special interest groups, 1
spill, of radioactive materials, 99, 110
statistics (probability of cancer), 28, 91, 102
steam
   engine, 41, 42, 67
   turbine, 41, 42, 60–74
stopping length or distance of radiation particles (alphas and betas), 97
storage battery, electric, 57–59
Strassman, Fritz (nuclear physicist, co-discoverer of nuclear fission), 44, 70
strontium-90 (Sr-90), fission product radioisotope, 99
subcritical, fissionable mass, 76, 123, 124
submarine nuclear reactors, 27, 110
sulfur dioxide (SO_2), air pollution by coal burning, 25, 88
sulfur dioxide, as catalyst in hydrogen production, 55
supercritical, fissionable mass, 6, 111, 113–114, 123–124
swimming-pool, storage of spent fuel, 82
synfuels (synthetic portable fuels)
   for automobile propulsion, 52–57, 61–67
   for aircraft (aviation), 53, 63
   for rockets and space vehicles, 57
synpetrol, 52, 54

tar-sands oil, 36, 37
TBP, TriButyl Phosphate for fuel reprocessing, 88
Teller, Edward (world famous pioneer molecular and nuclear physicist), 93

temperature coefficient of reactivity, 25, 76, 81, 113, 116
terrorists, 122–125
Tesla, Nikola (electric power pioneer – inventor AC motor, working for Westinghouse), 42
Texaco Oil Company, 43
thermal neutron, 70–76
thermalization (moderation) of neutrons, 70–76
thermal pollution, by power reactors, 76
Thomson, Elihu (electric power pioneer, working for Gneral Electric), 42
thorium-232, as breeding source for U-233, 79
thyroid, attraction for iodine, 99
Three-Mile-Island (TMI) reactor accident, 25–26, 112–113
tidal-wave, energy extraction from, 48
Tokaimura, nuclear supercritical accident, 111, 122
toxic chemicals, use by terrorists, 123, 124
trauma, from radiation overdose exposure, 106
tritium, nuclear fusion fuel, 93
turbogenerator, for electric energy generation, 42, 69–71

ultraviolet (UV) radiation, 106
ultracentrifuge (UCF), for uranium enrichment, 85–87
uranium
  consumption rate, 34–37
  depletion time of reserves, 22, 34–37
  earth reserves, 82, 34–37
uranium enrichment, 84–88
  AVLIS (atomic vapor laser isotope separation), 87
  DIF, gaseous diffusion method, 84–87
  Electromagnetic (Calutron), 84–85
  MLIS, molecular laser isotope separation method, 86–87
  UCF, ultracentrifuge method, 85–87
uranium hexafluoride gas, for molecular enrichment processes, 82, 83
uranium metal vapor, in AVLIS enrichment process, 87
uranium isotopes, 44, 70–73, 79, 149
URENCO enrichment corporation (Great Brittain, Germany, Netherlands), 86
USEC enrichment corporation (USA), 86
Utah, University of, cold-fusion experiments, 94

vacuum, high-, for uranium enrichment, 85–87
vapor, uranium, in AVLIS enrichment process, 87
victims, of Chernobyl accident, 27, 28, 114
vitrification of radiowaste, 88

WASH-1400, reactor safety study (Ref. 33), 90–92
waste, reactor heat, 10, 46
waste in energy generation
  coal, 25, 88
  oil and natgas, 8, 53, 63, 66
  uranium, 72, 79, 98
waste, radioactive, 26, 27, 82, 83, 88–92
waterfalls (hydro), as source of energy, 21, 48
water
  + air + electricity to make synfuel, 52ff
  as neutron moderator and reflector, 72–77
  as reactor coolant, 72–77
  as steam to run engines, 67
  as steam to run turbogenerators, 42, 70–77
Watt, James (18th century engineer), 41
weapons, nuclear, 19, 20, 46, 79–81, 125–129
Weiszäcker, von C.F. (early nuclear fusion (astro)physicist), 93
Westinghouse, George (AC electricity generation and distribution pioneer), 42
Wilson, E.D. (early nuclear fusion (astro)physicist), 93
wind energy, windmills, wind turbines (Refs 43, 44), 12, 13, 14, 21, 40, 41
world realities in nuclear age, 117–122
World Trade Center, terrorist attack, 124
Wright, Orville and Wilbur (engine-powered aircraft pioneers), 42

xray
  dose from dental/medical, 106
  photon radiation, 97, 99, 106

Yucca nuclear radiowaste repository, 10, 88–92

Z (Atomic charge number = number of protons in nucleus), 44, 72, 149

TOPICS IN SAFETY, RISK, RELIABILITY AND QUALITY

1. P. Sander and R. Badoux (eds.): *Bayesian Methods in Reliability*. 1991
   ISBN 0-7923-1414-X
2. M. Tichý: *Applied Methods of Structural Reliability*. 1993
   ISBN 0-7923-2349-1
3. K.K. Aggarwal: *Reliability Engineering*. 1993
   ISBN 0-7923-2524-9
4. G.E.G. Beroggi and W.A. Wallace (eds.): *Computer Supported Risk Management*. 1995
   ISBN 0-7923-3372-1
5. M. Nicolet-Monnier and A.V. Gheorghe: *Quantitative Risk Assessment of Hazardous Materials Transport Systems*. Rail, Road, Pipelines and Ship. 1996
   ISBN 0-7923-3923-1
6. A.V. Gheorghe and R. Mock: *Risk Engineering*. Bridging Risk Analysis with Stakeholders Value. 1999   ISBN 0-7923-5574-1
7. I.N. Vuchkov and L.N. Boyadjieva: *Quality Improvement with Design of Experiments*. A Response Surface Approach. 2001   ISBN 0-7923-6827-4
8. A.V. Gheorghe (ed.): *Integrated Risk and Vulnerability Management Assisted by Decision Support Systems*. Relevance and Impact on Governance. 2005   ISBN 1-4020-3451-2
9. A.V. Gheorghe, M. Masera, M. Weijnen and L. de Vries: *Critical Infrastructures at Risk*. Securing the European Electric Power System. 2006   ISBN 1-4020-4306-6
10. E.G. Frankel (ed.): *Challenging American Leadership*. Impact of National Quality on Risk of Losing Leadership. 2006   ISBN 1-4020-4892-0
11. Jeff W. Eerkens: *The Nuclear Imperative*. A Critical Look at the Approaching Energy Crisis. 2006   ISBN 1-4020-4930-7

springeronline.com